𝖈𝖑𝖆𝖗𝖊𝖓𝖉𝖔𝖓 𝖕𝖗𝖊𝖘𝖘 𝖘𝖊𝖗𝖎𝖊𝖘

INDUCTIVE LOGIC

FOWLER

a

London

HENRY FROWDE

New York

MACMILLAN AND CO.

3-28797?

THE ELEMENTS

OF

INDUCTIVE LOGIC

DESIGNED MAINLY

FOR THE USE OF STUDENTS IN THE UNIVERSITIES

BY

THOMAS FOWLER, D.D.

President of Corpus Christi College
Wykeham Professor of Logic in the University of Oxford
And Honorary Doctor of Laws in the University of Edinburgh

FIFTH EDITION

CORRECTED AND REVISED

Oxford

AT THE CLARENDON PRESS

M DCCC LXXXIX

PREFACE TO THE FIRST EDITION.

THE object of the following work is to serve as an introduction to that branch of scientific method which is known as Induction. It is designed mainly for the use of those who have not time or opportunity to consult larger works, or who require some preliminary knowledge before they can profitably enter upon the study of them.

To the works of Mr. Mill, Dr. Whewell, and Sir John Herschel, the Author must, once for all, express his obligations. 'He has, however,' if he may be allowed to repeat the language already employed in the Preface to his *Manual of Deductive Logic,* 'endeavoured, on all disputed points, to reason out his own conclusions, feeling assured that no manual, however elementary, can be of real service to the student, unless it express what may be called the "reasoned opinions" of its author.'

The analysis of Induction presents far more difficulties than that of Deduction, and requires to be illustrated by far more numerous and intricate examples. But, on the other hand, it is more interesting both to the teacher and to the student; and, being a comparatively recent study, is less hampered by conventionalities of treatment. Since the time of Bacon, it has always, with more or less of success, claimed a place in liberal education, and many, to whom the technical terms and subtle distinctions of the older logic are justly repulsive, have experienced a peculiar delight in attempting to discover and test the grounds on which the results of modern science mainly rest.

The study of Deductive Logic can be of little service unless it be supplemented by, at least, some knowledge of the principles of Induction, which supplies its premisses. Many of the objections directed against the study of Logic are due to the narrow conceptions which are entertained of its province, and might be easily met by showing that the study, when we include both its parts, has a much wider range than is popularly assigned to it.

Though the present work is mainly intended for students in the Universities, it is hoped that it will be found to present some interest for the general reader, and that it may be useful to students of medicine and the physical sciences, as well as to some of the more advanced scholars in our Public Schools.

The number of scientific examples adduced throughout the work renders it necessary, perhaps, that the Author should state emphatically that the work is intended as an introduction, not to *science*, but to *scientific method*. Its object is not to give a résumé of the sciences, physical or social, a task for which the Author would be wholly incompetent, but to show the grounds on which our scientific knowledge rests, the methods by which it has been built up, and the defects from which it must be free. Notwithstanding its frequent incursions into the domain of science, the purport of the work must be regarded as strictly logical.

The examples have, as a rule, been selected from the physical rather than the social sciences, as being usually less open to dispute, and lying within a smaller compass. Wherever it has been possible, they have been given in the exact words of the author from whom they are taken.

Some of the more complicated cases of inductive reasoning, such as those, which deal with Progressive Causes or Intermixture of Effects, have, if alluded to at all, been only briefly noticed. Any detailed examination of these more intricate questions seemed to lie without the scope of the treatise. The student who has leisure to pursue the subject will find ample information in the pages of Mr. Mill's *Logic*.

It only remains for the Author to express his grateful acknowledgments to those who have assisted him in the execution of the work. These are, in the first place, due to Dr. Liddell, Dean of Christ Church, through whose hands the sheets have passed, and who, in addition to revising the proofs, has, from time to time, offered many very valuable suggestions. They are due also, in no small degree, to Sir John Herschel and Professor Bartholomew Price, who most kindly undertook to revise the scientific examples; to Professors Rolleston and Clifton, who have frequently allowed the Author to consult them on questions connected with the subjects of their respective chairs, and to the Rev. G. W. Kitchin, the Organising Secretary of the Clarendon Press Series. The Author must, however, be regarded as alone responsible for any errors which may occur either in the theoretical portion of the work or in the examples.

LINCOLN COLLEGE,
Oct. 30, 1869.

PREFACE TO THE THIRD EDITION.

[The student is requested to read this Preface in connexion
with Chapter III.]

SINCE the publication of my second Edition, there has
appeared an important work on Scientific Method, entitled
'The Principles of Science,' by Professor Stanley Jevons,
of Owens College, Manchester. To this book I have
made occasional references in the foot-notes to my present
edition. But, as I differ entirely from Professor Jevons
on the fundamental question of the validity of our inductive inferences, I think it desirable to offer a few remarks
on this point in the present place, rather than to introduce controversial matter into the body of the work.

Mr. Jevons over and over again asserts the uncertainty,
or the mere probability, of all inductive inferences. Thus,
for instance, in his chapter on the Philosophy of Inductive Inference, he says :—'I have no objection to use
the words cause and causation, provided they are never
allowed to lead us to imagine that our knowledge of
nature can attain to certainty[1].' And again : 'We can
never recur too often to the truth that our knowledge of

[1] Vol. i. p. 260.

the laws and future events of the external world is only probable[2].'   Once more: 'By induction we gain no certain knowledge; but by observation, and the inverse use of deductive reasoning, we estimate the probability that an event which has occurred was preceded by conditions of specified character, or that such conditions will be followed by the event[3].'

At the same time, I am quite unable to reconcile with these passages other passages, such as those in which Mr. Jevons says: 'We know that a penny thrown into the air will certainly fall upon a flat side, so that either the head or tail will be uppermost[4],' or, 'I can be certain that nitric acid will not dissolve gold, provided I know that the substances employed really correspond to those on which I tried the experiment previously[5].'

But, waiving the question of inconsistency, I maintain as against Mr. Jevons that many of our inductive inferences have all the certainty of which human knowledge is capable. Is the law of gravitation one whit less certain than the conclusion of the 47th Proposition of the First Book of Euclid? Or is the proposition that animal and vegetable life cannot exist without moisture one whit less certain than the truths of the multiplication table? Both these physical generalisations are established by the

[2] Vol. i. p. 271.   [3] Id. p. 257.

[4] Id. p. 228. Mr. Jevons, however, curiously enough is not certain about the truth of the Law of Gravitation. See below.

[5] Id. p. 270.

Method of Difference, and, as *actual* Laws of Nature, admit, I conceive, of no doubt. But it may be asked if they will always continue to be Laws of Nature? I reply that, unless the constitution of the Universe shall be changed to an extent which I cannot now even conceive, they will so continue, and that no reasonable man has any practical doubt as to their continuance. And why? Because they are confirmed by the whole of our own experiences, which in both these cases is of enormous extent and variety, by the experience of our ancestors, and by all that we can ascertain of the past history of nature, while their reversal would involve the reversal of almost all the other laws with which we are acquainted. Still, it must be confessed that all our inferences from the present to the future are, in one sense, hypothetical, the hypothesis being that the circumstances on which the laws themselves depend will continue to be the same as now, that is, in the present case, that the constitution of nature, in its most general features, will remain unchanged; or, to put it in still another form, that the same causes will continue to produce the same effects. What would happen if this expectation were ever frustrated, it is absolutely impossible for us to say, so completely is it assumed in all our plans and reasonings.

We may say, then, that there are many inductions as to the *actual* constitution of nature which we may accept with certainty, while, with respect even to the distant future, we may accept them with equal certainty, on the hypothesis that the general course of nature will not be radically changed. And if the general course of nature were changed, might not the change affect our faculties as well as the objects of our knowledge; and, in that case, are we certain that we should still regard things that are equal to the same thing as equal to one another, or assume that a thing cannot both be and not be in the same place at the same time? There is, in fact, no limit to the possibility of scepticism with regard to the persistency either of the laws of external nature or of the laws of mind. But all our reasonings depend on the hypothesis that the most general laws of matter and the most general laws of mind will continue to be what they are, and of the truth of this hypothesis no reasonable man entertains any practical doubt[6].

There is, then, I contend, no special uncertainty attaching to the truths arrived at by induction. They are, indeed, like all other truths, relative to the present constitution of nature and the present constitution of the human mind, but this is a limitation to which all our knowledge alike is subject, and which it is vain for us to attempt to transcend. Syllogistic reasoning implies a particular constitution of the mind, as much as induc-

---

[6] Thus Mr. Jevons, who, when he begins to theorise, has doubts as to the truth of the Law of Gravitation, has no doubt, when he throws a penny up into the air, that it will fall on a flat side.

tive reasoning implies a particular constitution of nature. Both mind and nature might, of course, be radically changed by an omnipotent power, but what the consequences of that change might be it is utterly impossible for us to say.

The uniformity of nature, the trustworthiness of our own faculties—these are the ultimate generalisations which lie at the root of all our beliefs, and are the conditions of all our reasonings. It is, of course, always possible to insinuate doubts as to either, but, however curious and entertaining such doubts may be, they have no practical influence even on those who originate them. Even Mr. Jevons himself, we have seen, when not under the dominion of his theory, speaks of some of the results of induction as certain, and we can hardly conceive men of science commonly speaking of the most firmly established generalisations of mechanics, optics, or chemistry, simply as conclusions possessing a high degree of probability.

Still, Mr. Jevons, appearing not in the character of a physicist, but of a logician, tells us that 'the law of gravitation itself is only probably true[7].' It would be interesting to learn what is the exact amount of this 'probability,' or, if it be meant that we can only be certain that the force of gravity is acting here and now, it would be an interesting enquiry to ascertain what is

[7] p. 300.

the exact value of the 'probability' that it is at this moment acting in Manchester as well as in Oxford, or that it will be acting at this time to-morrow as well as to-day.

But, if the conclusions of Induction are thus uncertain, where, according to Mr. Jevons, are we to find certainty? 'Certainty belongs only to the deductive process and to the teachings of direct intuition[8].' Does it then belong to the conclusions of deduction? Apparently not, for, at the very beginning of the work[9], we are told that 'in its ultimate origin or foundation, all knowledge is inductive,' and Mr. Jevons is, of course, too practised a logician to suppose that the conclusion can be more certain than the premisses. The conclusions of geometry, therefore, partake of the same 'uncertainty' as the results of the physical sciences, and the region of 'certainty' is confined to our direct intuitions and to the rules of syllogism (supposing, that is, a difference to be intended between the 'deductive process' and deductive results). I venture to suggest that this small residuum of 'certainty' would soon yield to solvents as powerful as those which Mr. Jevons has applied to the results of induction (and apparently also of deduction); and that, therefore, its inherent 'uncertainty' is no special characteristic of that method, but one which it shares with all our so-called knowledge.

[8] p. 309.   [9] p. 14.

The fact is that in all reasoning, whether inductive or deductive, we make, and must make, assumptions which may theoretically be questioned, but of the truth of which no man, in practice, entertains the slightest doubt. Thus, in syllogistic reasoning, we assume at every step the trustworthiness of memory; we assume, moreover, the validity of the premisses, which, as Mr. Jevons acknowledges, must ultimately be guaranteed either by induction or direct observation; lastly, we assume the validity of the primary axioms of reasoning, which, according to different theories, are either obtained by induction or assumed to be necessary laws of the human mind. In this sense, all reasoning and all science is hypothetical, and the assumption of the Uniformity of Nature does not render inductive reasoning hypothetical in any special sense of the term. For, if the Laws of the Uniformity of Nature and of Universal Causation admit of exceptions or are liable to ultimate frustration, so, for aught we know, may the axioms of syllogistic reasoning or the inductions by which we have established the trustworthiness of our faculties. And, if the conceptions of uniformity and causation be purely relative to man, so, for aught we know, may be the so-called laws of thought themselves[10]. Induction

[10] According to the view of the nature and ultimate origin of human knowledge, accepted both by Mr. Jevons and myself, it is, in fact, no paradox but a mere truism to say that the fundamental axioms of reasoning are themselves only particular uniformities of

would only be hypothetical in a special sense, if we had any reasonable ground for doubting the truth of the hypotheses [11] on which it rests.

But as, 'in its ultimate origin or foundation, all knowledge' (including, of course, that of the laws which govern the syllogistic process itself) 'is inductive,' Professor Jevons must either employ the word 'certain' in a variety of senses, or he must be prepared with the philosophers of the New Academy to maintain the uncertainty of all knowledge whatsoever.

Such, as it appears to me, are the inconsistencies and paradoxes into which a very able writer has been led by a tendency to over-refinement, and, still more perhaps, by a desire to apply the ideas and formulæ of mathematics to the explanation of logical problems.

I must further express my dissidence from Mr. Jevons' statement that all inductive inference is preceded by

nature, arrived at by the same evidence and depending for their justification on the same grounds as those ultimate generalisations on causation to which we give the special names of the Law of Universal Causation and the Law of the Uniformity of Nature.

[11] I need hardly say that I am not here using the word 'hypothesis' in the sense of an unverified assumption. Reasoning, both inductive and deductive, is found on analysis to depend, in the last resort, on certain assumptions or hypotheses, but then the truth of these assumptions or hypotheses is guaranteed by the whole experience of the human race, past and present, and beyond this guarantee I conceive that there is no other attainable. In other words, all truth is relative to our faculties of knowing, and this condition it is in vain for us to attempt to transcend.

hypotheses [12], from his theory that Induction is simply the Inverse Method of Deduction, and, above all, from what appears to me to be the exceedingly misleading parallel drawn between Nature and a ballot-box. 'Events,' says Mr. Jevons, 'come out like balls from the vast ballot-box of Nature [13].' Now the balls were placed in the ballot-box by human hands; the number and character of them may have been due merely to caprice or chance; moreover, they are all isolated entities having no connexion with each other. Would it be possible to find a stronger contrast to the works of nature? If natural phenomena did indeed admit only of the same kind of study as the drawing of balls from a ballot-box, Mr. Jevons' conception of Induction would undoubtedly be the true one, and I should agree with him that 'no finite number of particular verifications of a supposed law will render that law certain.' But, just because we believe that the operations of Nature are conducted with an uniformity for which we seek in vain amongst the contrivances of men, do we regard ourselves as capable, in many cases, of predicting the one class of events with certainty, while the other affords only matter for more or less probable conjecture.

Intimately connected with Mr. Jevons' depreciation of the value of the inductive inference is his statement that

[12] See chap. i. pp. 11–13, of this work. [13] Vol. i. p. 275.

Induction is simply the inverse method of Deduction. If Induction simply consists in framing hypotheses, deducing consequences from the hypotheses, and then comparing these consequences with individual facts for the purpose of verifying them by specific experience [14], I grant that the procedure must, in most cases, be very untrustworthy. In my first Appended Note to my Section on Hypothesis, I have examined this account of Induction, which is virtually identical with that of Dr. Whewell. In opposition to it, I maintain the following theses, which are explained and defended in the course of my work: 1. That our inductions are not always preceded by hypotheses (and it might be added that, even where they are, the hypothesis itself must rest originally on some basis of fact, that is to say, on some induction or other, however imperfect; for a hypothesis must always be suggested by something of which we have had experience); 2. That the mere verification of our hypotheses by specific experience is not sufficient to constitute a valid induction, unless the instances conform with the requirements of one of the inductive methods, or (as in the case of the fundamental laws of inductive reasoning) be coextensive with the whole experience of mankind. Induction, I maintain, may or may not employ hypothesis, but what is essential to it is the inference from the particular

[14] Vol. i. pp. 307, 308.

to·the general, from the known to the unknown, and the nature of this inference it is impossible to represent adequately by reference to the forms of deduction [15].

[15] For the word ' adequately,' I ought to substitute the expression, ' without a considerable amount of circumlocution,' as the essential difference between inductive and deductive reasoning consists, not so much in the form of the argument, as in the nature of the assumptions made: scientific induction postulating, in addition to the assumptions made in deductive reasoning, the laws of Universal Causation and of Uniformity of Nature, in its strictest sense (see p. 9, note 7); and Inductio per Enumerationem Simplicem, the latter law in its vaguer sense. I think it may be useful to the student here to transcribe (with some subsequent modifications) a note which appeared in the last (9th) edition of my Deductive Logic (pp. 73, 74, note 3).

'If we state explicitly all the assumptions made in the inductive process, the conclusion is contained in the premisses, and the form of the reasoning becomes deductive; but it is seldom that we do state our assumptions thus explicitly. The most essential distinction, however, between inductive and deductive reasoning consists not in the form of the inferences, but in the nature of the assumptions on which they rest. Deductive reasoning rests on certain assumptions with regard to language and co-existence (namely, the Law of Contradiction, the Law of Excluded Middle, and the Canons of Syllogism), while inductive reasoning assumes over and besides these laws the truth of the Laws of Universal Causation, of the Uniformity of Nature and, as implied in the latter, of the Conservation of Energy; or, if it be of the unscientific description which is known as Inductio per Enumerationem Simplicem, it merely assumes, instead of them, the vague and wide principle that the unknown resembles, or will resemble, the known. It hardly needs to be added that all reasoning alike assumes the trustworthiness of present consciousness and of memory.

Amongst the assumptions or pre-suppositions of reasoning, I have not included the so-called Law of Identity; as to say that all·A is A,

Mr. Jevons' statement that 'induction is really the reverse process of deduction' I am wholly unable to reconcile with the following statements which occur in the very same page [16]: 'In its ultimate origin or foundation all knowledge is inductive,' and 'only when we possess such knowledge, in the form of general propositions and natural laws, can we usefully apply the reverse process of deduction to ascertain the exact information required at any moment.' When we compare these statements, the circle seems complete. A precedes B, and B precedes A. A depends for its validity on B, and B depends for its validity on A. No wonder that human reasoning affords us no 'certain' results.

In offering these criticisms on some fundamental points of difference between Mr. Jevons and myself, I am far from denying the utility of many portions of his work, especially the chapters on the Methods of Measurement and on Hypothesis.

In the present Edition of this work, I have occasionally availed myself of the 'Inductive Logic' of

or a thing is the same as itself, appears to me to be an utterly unmeaning proposition. Mr. Mill (Examination of Hamilton, ch. 21), in attempting to give a meaning to this maxim, really transforms it into a perfectly distinct proposition, namely, that Language may express the same idea in different forms of words.'

[16] Vol. i. p. 14.

Mr. Bain, a work which, though it does not, in my opinion, supersede Mr. Mill's Logic, supplies on some points a valuable complement to it. ·

In this, as in the last Edition, I have to acknowledge the kindness of Professor Park of Belfast, whose corrections and suggestions have enabled me to make both my works more accurate and serviceable than they would otherwise have been.

LINCOLN COLLEGE,
Feb. 24, 1876.

_{*} In the third Edition some new matter was introduced, bearing mainly on the following subjects: Uniformities of Coexistence, the Historical· Method, the distinction between Inductio per Enumerationem Simplicem and the Method of Agreement, the constant alternation in practice of the inductive and deductive processes, and the Argument from Universal Consent. In the fourth Edition the principal alterations were the introduction of new foot-notes on the definition of Induction and on the Plurality of Causes, and some additional remarks on the nature of the Method of Residues and on Empirical Laws.

In the fifth Edition, the alterations are more numerous than in either of the two preceding editions. Through the kindness of my friend, Mr. George Griffith of Harrow, I have been enabled to state some of the scientific examples in a more precise form than in the preceding editions, notably those on Double-Weighing (p. 47), on the 'Red Flames' seen during a total eclipse of the Sun (pp. 50–1), and on Spectrum Analysis as applied to the constitution of the Sun and other heavenly bodies (pp. 165–7). The principal alterations or additions in the logical matter are on pp. 204–5 (the Historical Method), p. 206 (the Comparative Method), pp. 227–8 (the Fallacy of 'Exaggerated Comparison'), and the addition of an important foot-note (note 15, p. xix) to the 'Preface to the Third Edition,' on the peculiar nature of Inductive Reasoning and on the assumptions made in it. The following foot-notes are either new or contain additional matter: n. 4, p. 6; n. 11, p. 13; n. 22, p. 23; n. 41, pp. 107–8; n. 5, p. 129; n. 27, p. 166; n. 44, p. 191; n. 54, p. 205; n. 55, pp. 206–7; n. 63, p. 214; n. 24, pp. 281–2; n. 26, p. 282; n. 29, p. 283; n. 80, p. 343.

C. C. C.
Dec. 1, 1888.

CONTENTS.

CONTENTS.

'Εκ προγινωσκομένων δὲ πᾶσα διδασκαλία, ὥσπερ καὶ ἐν τοῖς ἀναλυτικοῖς λέγομεν· ἡ μὲν γὰρ δι' ἐπαγωγῆς, ἡ δὲ συλλογισμῷ. Ἡ μὲν δὴ ἐπαγωγὴ ἀρχή ἐστι καὶ τοῦ καθόλου, ὁ δὲ συλλογισμὸς ἐκ τῶν καθόλου. Εἰσὶν ἄρα ἀρχαὶ ἐξ ὧν ὁ συλλογισμὸς, ὧν οὐκ ἔστι συλλογισμός· ἐπαγωγὴ ἄρα.

Aristotle's Nicomachean Ethics, vi. 3 (3).

Quamvis ad scientiam quamlibet via unica pateat, quâ nempe a notioribus ad minus nota et a manifestis ad obscuriorum notitiam progredimur, atque universalia nobis præcipuè nota sint (ab universalibus enim ad particularia ratiocinando oritur scientia), ipsa tamen universalium in intellectu comprehensio a singularium in sensibus nostris perceptione exsurgit.

Preface to Harvey's Treatise *De Generatione Animalium.*

ELEMENTS

OF

INDUCTIVE LOGIC.

*** THE notes appended to the Chapters (as distinguished from the foot-notes) are designed to inform the student of any divergences from the ordinary mode of treatment, or to afford him information on disputed questions which it appeared inconvenient to notice in the text. They may be omitted on the first reading.

CHAPTER I.

On the Nature of Inductive Inference.

TWO bodies of unequal weight (say a guinea and a feather) are placed at the same height under the exhausted receiver of an air-pump. When released, they are observed to reach the bottom of the vessel at the same instant of time, or, in other words, to fall in equal times. From this fact, it is inferred that a repetition of the experiment either with these two bodies or with any other bodies would be attended with the same result, and that, if it were not for the resistance of the atmosphere and other impeding circumstances, all bodies, whatever their weight, would fall through equal vertical spaces in equal times. Now, that these two bodies in this particular experiment fall to the bottom of the receiver in equal times is merely a fact of observation, but that they would do so if we repeated the experiment, or that the next two bodies we selected, or any bodies, or all bodies, would do so, is an inference, and is an inference of that particular character which is called an Inductive Inference or an Induction[1].

What assumptions underlie this inference, and on what grounds does it rest?

[1] The student must throughout bear in mind the ambiguous use of the words Induction, Inference, &c., as signifying both the *result* and the *process* by which the result is arrived at. See *Deductive Logic*, Preface, and Part III. ch. i. note I.

My object in placing the two bodies under the receiver was obviously to answer a question which I had previously addressed to myself: viz. whether, when subject to the action of gravity[2] only, they would fall in equal or in unequal times. By exhausting the air in the receiver, I am able to *isolate the phenomenon*, and thus, by removing all circumstances affecting the bodies, except the action of gravity, to watch the effect of this cause operating alone. But in trying this experiment, in isolating the phenomenon, and asking what will be the effect of the action of gravity operating alone, I am evidently assuming that the effect, whatever it may be, will be entirely due to the cause or causes which are then and there in action; in other words, I am assuming that nothing can happen without a cause, that no change can take place without being preceded or attended by circumstances which, if we were fully acquainted with them, would fully account for that change. This assumption (which may be called *the Law of Universal Causation*) is universally admitted by mankind, or at least by the reflecting portion of mankind, though the grounds on which it is admitted have been variously stated; some justifying it by an appeal to the continuous and uncontradicted experience not only of the individual himself

---

[2] When I employ the expression 'action of gravity' or 'force of gravity,' I must not be understood as adopting any particular theory on the nature of the phenomenon which we call 'gravitation.' I use these terms simply because they are short and recognised phrases for expressing the fact that all terrestrial bodies, when left entirely free, fall in the direction of the earth's centre.

but of the human race, others by an appeal to the necessities of thought.

Thus far, however, we have only ascertained that the fact of these two particular bodies, in this particular instance, falling to the ground in equal times is due to the action of gravity, unimpeded by any other circumstances. But why should I infer that they, if the experiment were repeated, or any other two bodies, if exposed to the same circumstances, would behave in the same way? It is not enough to feel assured that nothing can happen without a cause, and that the only cause operating in this particular instance is the action of gravity; I must also feel assured that the same cause will [3] invariably be followed by the same effect, or, to speak more accurately, that the same cause or combination of causes, will, if unimpeded by the action of any other cause or combination of causes, be invariably followed by the same effect or combination of effects, or, to state the same proposition in somewhat different language, that, whenever the same antecedents, and none others, are introduced, the same consequents will invariably follow. This assumption (or law) is, like the former, universally admitted by mankind, or the reflecting portion of mankind, though the grounds on which it is admitted have been variously stated, some, as in the case of the former law, referring it to experience, others to certain necessities of thought arising

[3] The expression 'will' is used for the sake of brevity. The argument, however, is not simply from the present to the future, but from cases within the range of our experience to all cases, past, present, or future, without that range. See p. 32, note 31.

from the original constitution of the human mind. This
law may be called *the Law of the Uniformity of Nature*[4].

The argument, then, in the case which we have taken
as our instance, may be represented as follows :—

> I *observe* that these two bodies (though of unequal
> weight) reach the bottom of the receiver at the
> same moment.
>
> This fact must be due to some cause or com-
> bination of causes (Law of Universal Causation).
>
> The only cause operating in this instance is the
> action of gravity.
>
> ∴ The fact that these two bodies reach the bottom of
> the receiver at the same moment is due to the
> action of gravity, operating alone.
>
> But, whenever the same cause or combination of
> causes is in operation, and that only, the same
> effect will invariably follow (Law of Uniformity of
> Nature).

---

[4] It is, perhaps, necessary thus early to warn the student that the
converse of the Law of the Uniformity of Nature does not hold true.
Though the same cause, that is, the same antecedent or combination
of antecedents, is never followed by different effects, the same effect,
or, more strictly speaking (see pp. 127, 8), the same portion of an
effect, may be due to different causes. We can, thus, always argue
from the cause to the effect, but we cannot always argue from the
effect to the cause.

The Law of the Uniformity of Nature implies, I conceive, the
truth of the law of the Persistence or Conservation of Energy
(namely, that no cause, or part of a cause, will ever be ineffective,
or, in other words, that no energy is ever lost), and hence I have not
thought it necessary to introduce any express mention of this latter
law in the text.

.·. Whenever these two bodies, or any other two or more bodies (even though of unequal weight), are subject to the action of gravity only, they will reach the bottom of the receiver at the same moment, or, in other words, will fall in equal times.

The induction just examined has been arrived at by a process of elimination, and takes for granted the conception of causation. It is representative of the inductions with which science is mainly concerned, and of which I shall have, for the most part, to treat in the present work. But there are other inductions of a simpler character, the validity of which is assured not by any artificial process of elimination, but merely by a series of uncontradicted experiences. This kind of induction is usually distinguished by logicians as *Inductio per Enumerationem Simplicem*. It is often (as will hereafter be pointed out in the 4th chapter) exceedingly untrustworthy, but, when the area of experience is very wide, the evidence which it affords may approach to, and even amount to, certainty. Often moreover, and especially in the case of our widest generalisations, it is our only resource.

Amongst inductions of this kind must be included, as I conceive, the Laws of Uniformity of Nature and Universal Causation themselves, as well as the axioms of mathematics and certain facts of co-existence which have not yet been resolved into, or possibly do not admit of being resolved into, facts of causation. As examples of the last class I may specify the co-existence throughout matter of the properties of inertia and gravity, and the co-inherence of attributes in the various kinds of animals, plants, and minerals, as, for instance, fusibility at a certain point together with a certain specific gravity in gold, or the combination of rationality with a peculiar physical form in man. Though co-existing facts of this nature may possibly be due to some causal connexion, and might, if we had a perfect knowledge of all natural processes, be explained in that manner, they are, as yet, known to us only as facts of co-existence, and established only by an inductio per enumerationem simplicem, or uncontradicted experience.

Mr. Bain[5] enumerates three kinds of uniformities, which may be established by induction, those of *Co-existence*, *Causation*, and *Equality*. Uniformities of Co-existence and Equality can be established only by Inductio per Enumerationem Simplicem, while those of Causation, though, in the actual state of our knowledge, they often rest only on this evidence, ought always to be established by the more refined methods to be described in the sequel of this book. To the above classification I ought to add the Laws of Uniformity of Nature and Universal Causation, both of which, as already remarked, I conceive to be established by uncontradicted experience, or, in other words, by an Inductio per enumerationem simplicem coextensive with all human knowledge[6]. These fundamental laws, thus verified by a constant experience,

[5] *Logic*, Bk. III. ch. ii.

[6] On the nature of the evidence on which these laws rest, see the third appended note at the end of this chapter.

are assumed in all scientific inductions concerning Causation, whereas, in mere inductions *per Enumerationem Simplicem*, all that is assumed is the much vaguer and less precise belief that, under similar circumstances, the unknown resembles, or will resemble, the known, a belief which experience shows to be subject to many modifications [7].

As the inductions of Causation are those with which science is mainly concerned, and to which alone the more refined rules of Inductive Logic are applicable, I shall in the following work limit myself almost entirely to their consideration. The inductions of Co-existence, with which I shall, to some extent, be concerned in the section on Classification, I shall regard as subservient to these [8].

From what has been said above, as well as in distinguishing the various kinds of inference in the *Manual of Deductive Logic*, it will be seen that Induction may be defined as *the legitimate inference of the unknown from the known*, that is, of propositions applicable to cases hitherto unobserved and unexamined from propositions which are known to be true of the cases observed and examined. Thus, from the proposition that a guinea

---

[7] This belief is what is frequently understood by the Law of the Uniformity of Nature, but I have thought it desirable to confine that expression to the more precise statement with regard to the uniform action of causes.

[8] I shall briefly recur to the subject of Inductions of Co-existence in the fourth chapter, under the head of *Inductio per Enumerationem Simplicem*.

and a feather, if placed under the exhausted receiver of an air-pump, will fall through equal vertical spaces in equal times, may be inferred inductively the proposition that a shilling, a penny, and a straw will, if exposed to the same circumstances, also fall in equal times. But, as we can only draw this inference on grounds which are equally applicable to all bodies whatsoever, when exposed to the same circumstances, and as we might make the same assertion of *any* two or more bodies, and consequently of *all* bodies, it will be seen that Induction is not only an inference of the unknown from the known ; but, in virtue of that fact, of the general from the particular. In every inductive argument, in fact, it is implied that *wherever* or *whenever* the same circumstances are repeated, the same effects will follow. Induction may, therefore, also be defined as *the legitimate inference of the general from the particular*, or (in order to include those cases where general propositions are themselves employed as the starting-point of an inductive argument, of which numerous instances will occur as we proceed) *of the more general from the less general* [9].

[9] This is a better and more accurate definition than that given at the beginning of the paragraph, because, if we adopt the theory that all our fundamental beliefs are derived from experience, there is no kind of inference which does not involve the assumption that we may argue from the known to the unknown, and from the past to the present and the future. In the case of some of these beliefs, however, as the so-called 'laws of thought,' and the belief in the trustworthiness of our present consciousness and of memory, the assumption has been made so often and so constantly that we have almost ceased to be conscious of making it.

In trying the experiment which furnished our instance at the beginning of the chapter, we were attempting to find an answer to the question, 'Do bodies, when subject to the action of gravity only, fall through equal vertical spaces in equal or in unequal times?' The experiment may be regarded as an attempt to decide between two rival theories (or *hypotheses*, as they are usually called), one being that bodies fall quicker in proportion to their weights, the other that the weight of the body, when the resistance of the atmosphere is removed, does not affect the time of falling. The experiment is decisive in favour of the latter hypothesis, which is thus entitled to rank as a valid induction. Our inductions are often, as in this case, the result of an attempt to decide between rival hypotheses, or a reply to the question whether some particular hypothesis be true or not, the hypothesis or hypotheses suggesting the particular experiment to be tried. Sometimes, however, we have no assistance of this kind, and we try experiments simply 'to see what will come of them.' Thus, if a chemist discovers a new  element, he will proceed to try a variety of experiments in order to determine the proportions in which it will combine with other elements, as well as to discover the various properties of such combinations. Supposing the experiments to have been properly conducted, the inductions at which he arrives will be perfectly valid, though he may have formed no previous theories as to the results of his researches. Occasionally, too, an induction will not be the result of any definite course of investigation, but will be obtruded on our notice, as in the following instance, adduced by Sir John Herschel, to show that 'after much labour in vain, and groping in the dark, accident or casual observation will present a case which strikes us at once with a full insight into a subject.' 'The laws of crystallography were obscure, and its causes still more so, till Haüy fortunately dropped a beautiful crystal of calcareous spar on a stone pavement, and broke it. In piecing together the fragments, he observed their facets not to correspond with those of the crystal in its entire state, but to belong to another form ; and following out the hint .... thus casually obtruded on his notice, he discovered the beautiful laws of the cleavage, and the primitive forms of minerals [10].'

Thus, we perceive that our inductions are sometimes preceded by hypotheses, at other times not. In most cases, probably, we have formed some supposition (or hypothesis) as to the character of a phenomenon before we enter upon, or, at least, before we complete, its investigation. Such suppositions (or hypotheses) are often of the utmost service in directing the course which our experiments and observations shall take. Frequently, also, it is impossible to perform any experiment, or to institute any series of observations, which shall be decisive of the question before us. In this case, unless we altogether suspend our judgment, we must rest content with an unproved theory, and it becomes of prime importance to determine to what conditions such a

[10] Herschel's *Discourse on the Study of Natural Philosophy*, § 191.

theory[11], supposition, or hypothesis must conform in order to entitle it to rank as a probable or possible solution of our difficulties. A subsequent section will be specially devoted to these questions, but meanwhile it seemed desirable at once to direct the attention of the student to the distinction between hypothesis and induction. He must bear in mind that, though the formation of hypotheses is frequently an important step in the inductive process, a hypothesis must be carefully distinguished from a valid induction. Without at present attempting any formal definition of a hypothesis, it may be distinguished from an induction (that is, a valid, complete, or perfect induction) as a mere supposition or assumption from an ascertained truth.

*** The word 'cause' is commonly used in a very vague and indefinite sense. Of the various antecedents whose presence or absence is essential to the event, it is usual to single out one as the Cause, and either to overlook the others, or to speak of them as 'conditions.' Strictly speaking, however, the Cause consists in the presence of all those antecedents, the withdrawal of any of which, and in the absence of all those antecedents, the introduction of any of which, might frustrate the event. Thus, to take the homely instance of lighting a fire.

[11] The word 'theory' is, unfortunately, employed in two meanings : (1) as = hypothesis, as when we speak of the undulatory theory, the Darwinian theory, or two or more 'rival theories'; (2) as an ascertained truth, or body of truths, as when we speak of the 'lunar theory,' or the 'theory of equations.'

The application of the lighted match is what would ordinarily be called the cause of the combustion. But there are other conditions equally necessary, as, for instance, amongst the positive conditions, the presence of fuel and of atmospheric air, and, amongst the negative conditions, the absence of such a quantity of moisture as would prevent the fuel from igniting. In assigning the cause of a phenomenon, it is seldom that the negative conditions are mentioned. It is generally understood that we assign a cause, subject to the qualification 'no counteracting cause intervening.' Amongst the positive conditions, we usually select that which, being last introduced, completes the assemblage of conditions, and stands in closest proximity to the effect. Thus, in our example, the combustion is said to be due to the application of the match, and, when a man, who has previously been in a bad state of health, is attacked by a fever, we speak of the fever as the cause of his death. These, however, as observed by Mr. Mill, are by no means invariable rules. 'It must not be supposed that in the employment of the term this or any other rule is always adhered to. Nothing can better show the absence of any scientific ground for the distinction between the cause of a phenomenon and its conditions, than the capricious manner in which we select from among the conditions that which we choose to denominate the cause. However numerous the conditions may be, there is hardly any of them which may not, according to the purpose of our immediate discourse, obtain that nominal pre-eminence.'. Thus, if a plot of dry heath is ignited by a spark from a railway-engine, we may, in common parlance, attribute the fire either to the spark, or to the dryness of the heath, or to the ill-construction of the engine; the first of these assigned causes being the proximate event, the second one of the other positive conditions, the last a negative condition. What, when employing popular language, we dignify with the name of Cause is that condition which happens to be most prominent in our minds at the time. It is, perhaps, superfluous to add that, when aiming at scientific accuracy, we ought to enumerate all the conditions, or, at least, all the positive conditions, on which a phenomenon depends, unless we have a right to presume that there is no likelihood of their being overlooked by those whom we address.[12]

In the science of Medicine, the cause which completes the assemblage of conditions is often distinguished as the *exciting* cause, the other causes being called *pre-disposing*. Thus, the peculiarities of constitution, age, sex, occupation, &c., which render a person more than ordinarily liable to any particular disorder, would be called the *pre-disposing* causes; the contagion (by which the body is brought into contact with some specific poison), a sudden chill, bodily fatigue, mental depression, or any circumstance, on the supervention of which the disease is immediately consequent, would be called

---

[12] The subject of this paragraph is treated with great ability in Mr. Mill's *Logic*, Bk. III. ch. v. § 3.

the *exciting* cause [13]. The pre-disposing causes of Asiatic Cholera, for instance, are enumerated in Dr. Guy's edition of Dr. Hooper's 'Vade Mecum,' as 'debility; impaired health; intemperance; impure air; low and damp situations; the summer and autumn seasons: the exciting causes as contagion; a peculiar poison diffused through the atmosphere.' The importance of attending to this distinction in historical and political investigations is forcibly stated and illustrated by Sir G. C. Lewis, in his *Methods of Observation and Reasoning in Politics*, vol. i. ch. ix. p. 333, &c.

Note 1 [14].—Mr. Mill (*Logic*, Book II. ch. iii.) maintains that, in an act of induction we usually, though not invariably, argue directly from one particular case to another. Dr. Whewell, on the other hand, holds that all inductive inference is from the particular to the general. (*Philosophy of Discovery*, ch. xxii. § 1–14.) Though I have adopted Dr. Whewell's language (which is that ordinarily employed), I cannot recognise the importance of the difference which he believes to exist between himself and Mr. Mill. To say that what I find to be true of this case will be true of the next which resembles it in certain assignable respects, whatever that case may be, or that what I found to be true of that case must be true of this, because this resembles that in certain as-

---

[13] See Dr. Watson's *Lectures on Physic*, Lecture VI.

[14] The student, unless he have some previous acquaintance with the subjects discussed in them, is recommended to omit these notes on the first reading.

signable respects, is virtually to say that it is true of any and every case which presents these points of resemblance. What is true of *each* or *any* case, taken indifferently, must be true of *all.* 'The burnt child dreads the fire.' Why? Because it once suffered pain, from burning its finger. Now, it appears to me indifferent whether we represent the child as having in its mind the proposition 'That object causes pain,' or the proposition 'That object will cause me pain now, if I approach too near to it.' But, as the former (the general) inference seems to be virtually implied in the latter (the particular), and, as Mr. Mill acknowledges, the particular inference can, on reflexion, only be justified by granting the truth of the general one, I prefer adhering to the common, and, as I think, the more intelligible account of induction. Mr. Mill himself, in one place, speaks of Induction as 'generalisation from experience,' and, in another, as 'the inference of a more general from less general propositions [15].'

Though agreeing with Dr. Whewell in his main position, I must express my entire dissent from the distinction which, throughout this discussion, he attempts to draw between our reasonings in the ordinary affairs

---

[15] Mr. Jevons (*Principles of Science*, vol. i. pp. 261, 262) seems to have slightly misapprehended my meaning in this note. While I believe that we do, as a matter of fact, often argue from particular to particular, I entirely agree with Mr. Jevons (*Principles of Science*, vol. ii. p. 243) in holding that 'what is inferred of a particular case might be inferred of all similar cases,' or, in other words, that the *logical justification* of such inferences is to be found in the general statement.

of life, and Induction as employed in scientific research. However various may be the conditions of their application, I cannot but regard the mental processes as identical, on whatever classes of objects they may be exercised. We may meet with insurmountable difficulties in the attempt to apply Induction to some obscure question of Physiology, and we may employ it with ease and . success a hundred times a day in compassing pleasure or avoiding pain, but I believe the mental process to be essentially the same in both cases.

Note 2.—Since the time of Hume, the nature of our conception of Cause has formed one of the chief topics of philosophical controversy. Previously to his . time, it appears to have been taken for granted by the great majority of modern philosophers of all schools [16] (if we except those who, like Malebranche, believed God to be the only efficient cause in the universe, and so-called acts of causation to be only the *occasions* of the Divine interference [17]), that the idea of causation necessarily implies the idea of *power* or *necessary connexion ;*

[16] Dugald Stewart (in his *Philosophy of the Human Mind*, Notes C and MM) has certainly succeeded in showing that Hume's views on the nature of Cause were anticipated by casual remarks of several other writers; but it still remains true that Hume was the first philosopher who definitely attacked the prevalent philosophical theory.

[17] Still, even according to these philosophers, every act of causation implies an act of power; only the power is exerted not by the so-called cause, but by the Deity himself. It will be noticed that I speak only of *modern* philosophers. Into the difficult question of the notions of causation entertained by ancient writers I do not enter.

necessary connexion, that is to say, between the cause and effect, or *power* in the cause to produce the effect. Even Locke, who effected a revolution in modern philosophy, left this idea of Power unassailed, though he attempted to account for its formation. 'The mind,' says he[18], 'being every day informed, by the senses, of the alteration of those simple ideas it observes in things without; and taking notice how one comes to an end, and ceases to be, and another begins to exist, which was not before; reflecting also on what passes within itself, and observing a constant change of its ideas, sometimes by the impression of outward objects on the senses, and sometimes by the determination of its own choice; and concluding from what it has so constantly observed to have been, that the like changes will for the future be made, in the same things, by like agents, and by the like ways, considers in one thing the possibility of having any of its simple ideas changed, and in another the possibility of making that change; and so comes by that idea which we call Power. Thus we say, fire has a power to melt gold, i. e. to destroy the consistency of its insensible parts, and consequently its hardness, and make it fluid; and gold has a power to be melted: that the sun has a power to blanch wax, and wax a power to be blanched by the sun, whereby the yellowness is destroyed, and whiteness made to exist in its room. In which, and the like cases, the power we consider is in reference to the change of perceivable ideas For we

[18] Locke's *Essay*, vol. ii. ch. xxi. § 1.

cannot observe any alteration to be made in, or operation upon anything, but by the observable change of its sensible ideas; nor conceive any alteration to be made, but by conceiving a change of some of its ideas.' He then proceeds to include our idea of Power amongst our Simple Ideas.

Hume contested the validity of this idea by an appeal to experience. Whence do we obtain this notion of necessary connexion between two events? Do we observe any such connexion in the events which take place in the external world, or in the relation between volition and the motion of the organs of the body, or in the act of the will by which it summons up, dwells on, or dismisses ideas? 'We have sought in vain for an idea of power or necessary connexion, in all the sources from which we could suppose it to be derived. It appears that, in single instances of the operation of bodies, we never can, by our utmost scrutiny, discover anything but one event following another; without being able to comprehend any force or power, by which the cause operates, or any connexion between it and its supposed effect. The same difficulty occurs in contemplating the operations of mind on body, where we observe the motion of the latter to follow upon the volition of the former, but are not able to observe or conceive the tie, which binds together the motion and volition, or the energy by which the mind produces this effect. The authority of the will over its own faculties and ideas is not a whit more comprehensible: so that, upon the whole, there appears not, throughout all nature, any one instance of connexion, which is conceivable by us. All events seem entirely loose and separate. One event follows another; but we never can observe any tie between them. They seem *conjoined*, but never *connected*. And as we can have no idea of anything, which never appeared to our outward sense or inward sentiment, the necessary conclusion *seems* to be, that we have no idea of connexion or power at all, and that these words are absolutely without any meaning, when employed either in philosophical reasonings, or common life [19].' Does Hume then deny the *fact of causation*, namely, that, when we have been accustomed to observe one event invariably followed by another, we may confidently expect, other circumstances remaining the same, that the one will continue to be followed by the other in the future, and that, if we perceive a change in any phenomenon, we may be confident that some other event has preceded that change? Certainly not. There is, in Hume's writings, absolutely no foundation for the virulence with which he is attacked by Reid [20]. What he called in

[19] Hume's *Essays.* Essay on the Idea of Necessary Causation.

[20] The following may serve as a specimen of Reid's diatribes against Hume. 'Of all the paradoxes this author has advanced, there is not one more shocking to the human understanding than this, That things may begin to exist without a cause. This would put an end to all speculation, as well as to all the business of life. The employment of speculative men, since the beginning of the world, has been to investigate the causes of things. What pity is it, they never thought of putting the previous question, Whether things have a cause or not? This question has at last been started; and

question was not the invariableness of the fact of causation, but the grounds of the prevalent notions attached to the word Cause. Whether his speculations on this subject be well or ill-founded, he certainly did not deny the correctness of the principles on which men act in ordinary life or which guide them in scientific research.

There is another objection to the statements contained in Hume's Essay which is better founded than the foregoing. If the term 'cause' be convertible with the term 'invariable antecedent,' it has been justly objected by Reid [21] that we might speak of day as the cause of night, and of night as the cause of day. That there are expressions in the Essay, in which the cause seems to be absolutely identified with the invariable antecedent or the sum of the invariable antecedents, cannot be denied. Such is the following: 'Suitably to this experience, what is there so ridiculous as not to be maintained by some philosopher?'—*Active Powers*, Essay I. ch. iv. Sir W. Hamilton and Dr. Mansel take a far juster view of Hume's position. Even Sir W. Hamilton, however, in commenting on Reid's statement, says, 'This' (namely, That things may begin to exist without a cause) 'is not Hume's assertion ; but that, on the psychological doctrine generally admitted, we have no valid assurance that they may not.' The latter is, certainly, not Hume's assertion. It is true that he bases the notion of causation on experience, but then he regards experience as the sole source of all our knowledge, other than that of mathematics. Sir William Hamilton's note requires only to be compared with the following passage from the Essay: 'But when one particular species of event has always, in all instances, been conjoined with another, we make no longer any scruple of foretelling one upon the appearance of the other, and of employing that reasoning which can alone assure us of any matter of fact or existence.'

[21] *Active Powers*, Essay IV. ch. iii.

therefore, we may define a cause to be an object, followed by another, and when all the objects, similar to the first, are followed by objects similar to the second.' But then the sentence proceeds: 'Or, in other words, *where, if the first object had not been, the second never had existed.'* Now this alternative definition is not open to Reid's objection [22], though it is open to the objection of ignoring the fact that the same event may be due to distinct causes, as pointed out in p. 6, n. 4 [23]. When modified to meet this objection, it would run thus: 'Cause [or causes] and Effect are two [or more] events, or sets of events, which are so related, that, if the first [or one of the first] had not been, the second had never existed.' Or, perhaps, it might be more simply stated thus: 'An Effect is so related to its Cause or its alternative Causes, that if the latter or one of the latter had not been, the former had never existed [24].'

[22] The alternative definition, however, introduces a new idea, not contained in the first definition, that of *dependence* of the effect upon the cause, which it is not easy to distinguish from the idea of necessary connexion. Hence, Hume appears unconsciously to recur to the very position which he is attacking. It seems to me that the relation of cause and effect, or the dependence of effect on cause, is an idea *sui generis,* and cannot be resolved into the mere idea of time, or antecedence and consequence. The introduction of the word 'power,' however, or even of the word 'necessary,' into the statement of the relation, occasions needless obscurity and difficulty.

[23] I am indebted to Professor Park of Belfast for drawing my attention to this objection, which had escaped my notice in the First Edition. It was originally pointed out by Dr. Thomas Brown, in his *Enquiry into the Relation of Cause and Effect*, Note A.

[24] This definition is, it must be confessed, somewhat deficient in

Both Dr. Thomas Brown and Mr. Mill attempt to meet Reid's objection.

Brown holds that the cause is that invariable antecedent which *immediately* precedes the effect; thus the position of the sun at a given moment, that which we call sun-rise, is the cause of day.

Mill attempts to meet the same objection by having recourse to the idea of '*unconditionalness.*' The cause of a phenomenon is 'the antecedent or concurrence of antecedents on which it is invariably and *unconditionally* consequent,' i.e. which not only invariably precedes it, but which is followed by the effect without the occurrence of any other condition. Now night cannot be called the cause of day, because it might go on for ever without being followed by day, unless the condition of the sun rising were fulfilled. See Mill's *Logic*, Bk. III. ch. v. § 5.

simplicity. But I venture to suggest that it will bear closer examination than that of Mr. Mill, who defines 'cause' as the 'unconditional invariable antecedent.' Whether by the term 'unconditional' he means 'not depending on any previous condition,' or 'not combined with any concurrent condition,' it may be objected that there is no such phenomenon in nature. If, as seems clear from the context, the term be used in the latter sense, we shall not only be excluded from saying that night is the cause of day, but also that solar light is the cause of day, for there are other conditions, both positive and negative, essential to the production of what we call day by the solar rays. In fact, according to the terms of this definition, we could never, as we are constantly doing, single out any one prominent phenomenon, and call it the cause of an event, without enumerating all the other conditions, positive and negative, which are essential to its operation.

The objection to Brown's account is that we fre quently speak without hesitation of A as the cause of B, though we are by no means certain that it is, strictly speaking, the *immediate* antecedent of B, nothing else whatever intervening; in fact, it is questionable whether in any case we can ascertain to a certainty that nothing else intervenes between two events. Similarly, it may be objected to Mill's account that we frequently speak without hesitation of A as being the cause of B, though we are by no means certain that there are not other antecedent conditions, positive and negative, which must be satisfied before A can be followed by B; and, indeed, as in the former case, it may be questioned whether we can ever be certain that there are no other conditions besides those which we have selected [25].

The first author of eminence who adopted Hume's view of the *nature* of Cause was Dr. Thomas Brown; singularly enough, however, so far from assuming with Hume that its *origin* was to be found in experience, he regarded it as instinctive. The notion of 'Power' he supposed was simply a gratuitous hypothesis, needlessly interpolated between the antecedent, which we call the Cause, and the consequent, which we call the Effect. 'We are eager to supply, by a little guess-work of fancy, the parts unobserved, and suppose deficiencies in our observation where there may truly have been none; till at length, by this habitual process, every phenomenon

[25] It is curious that Mill, in attempting to answer Reid, takes no notice of Brown's answer.

becomes, to our imagination, the sign of something intermediate as its cause, the discovery of which is to be an explanation of the phenomenon. The mere succession of one event to another appears, to us, very difficult to be conceived, because it wants that intervening something which we have learned to consider as a cause: but there seems to be no longer any mystery, if we can only suppose something intervening between them, and can thus succeed in doubling the difficulty, which we flatter ourselves with having removed; since, by the insertion of another link, we must now have two sequences of events instead of one simple sequence [26].' Hume's position is also accepted by James Mill in his *Analysis of the Phenomena of the Human Mind*, and by John Stuart Mill in his *System of Logic*.

Hume's antagonists have generally (with Kant) combated his arguments by denying the assumption on which they are based, namely, that the *origin* of our conception of Cause is to be sought in experience. Hume, it will be recollected, challenges those who maintain the hypothesis of 'power' or 'necessary connexion' to show how we can have become acquainted with it. Does it come from our experience of the external world, or from our experience of the control of our will over our own acts or our own thoughts? The answer of the Kantian School would be that it does not come from experience at all, that it is one of those fundamental conceptions which are native to the

[26] Brown's *Lectures on the Philosophy of the Human Mind*, Lecture IX. Cp. Lecture II.

human mind, not given by experience but evoked by it. Others, like Reid and Stewart, to whom we may add M. Maine de Biran, surrender the notion of power as applied to causation in the external world, while they maintain it as applied to our own actions, which are the results of will. We are conscious, they say, of power in ourselves, though we perceive only succession in the external world. Dr. Mansel, following Cousin, adopts a third view, and maintains that the notion of 'Power' is given only in the control of the mind over its own operations. 'The intuition of Power is not immediately given in the action of matter upon matter; nor yet can it be given in the action of matter upon mind, nor in that of mind upon matter; for to this day we are utterly ignorant how matter and mind operate upon each other. We know not how the material refractions of the eye are connected with the mental sensation of seeing, or how the determination of the will operates in bringing about the motion of the muscles. We can investigate severally the phenomena of matter and of mind, as we can examine severally the constitution of the earth, and the architecture of the heavens: we seek the boundary line of their junction, as the child chases the horizon, only to discover that it flies as we pursue it. There is thus no alternative, but either to abandon the inquiry after an immediate intuition of power, or to seek for it in *mind as determining its own modifications;*—a course open to those who admit an immediate consciousness of self, and to them only. My first and only presentation of power or causality is thus to be found in my consciousness of *myself as willing*[27].'

The relation subsisting between an act of will and the motion of the limbs, or between a physical antecedent and its consequent, he regards as beyond our knowledge. 'Our clearest notion of efficiency is that of a relation between two objects, similar to that which exists between ourselves and our volitions. But what relation can exist between the heat of fire and the melting of wax, similar to that between a conscious mind and its self-determinations? Or, if there is nothing precisely similar, can there be anything in any degree analogous? We cannot say that there is, or, if there is, how far the analogy extends, and how and where it fails. We can form no positive conception of a power of this kind: we can only say, that it is something different from the only power of which we are intuitively conscious. But, on the other hand, we are not warranted in denying the existence of anything of the kind; for denial is as much an act of positive thought as affirmation, and a negative idea furnishes no data for one or the other[28].'

It would, however, be beside my purpose to enter into a detailed account of the history of this controversy. In consequence, however, of its historical importance, it seemed essential to take some notice of it, and to point out that, whatever theory may be adopted as to the *nature* of Cause, and however great our inability to conceive *how* one event is followed by another, there is, at least, sufficient definiteness in the conception to entitle it to be accepted

[27] *Prolegomena Logica*, pp. 138, 139.   [28] *Ibid.* p. 140.

as the basis of scientific reasoning. Whether we acknowledge that one event has invariably the *power* of producing another, or whether we content ourselves with asserting that it is invariably followed by that other, it is, in either case, the element of *invariableness* which makes the connexion or conjunction, whichever we may call it, a fitting object of scientific research. But remove the element of invariableness, and suppose, if it be supposable, that the same antecedent or set of antecedents is sometimes followed by one consequent, and sometimes by another, and sometimes by none at all; in that case science would be impossible.

The student who wishes to obtain further information on this controversy (a controversy, however, which possesses a historical rather than a practical or scientific interest) is referred to Hume's *Essay on the Idea of Necessary Connexion;* Dugald Stewart's *Dissertation*, Part II. sect. 8; Mill's *Logic*, Book III. ch. v; Sir W. Hamilton's *Lectures on Metaphysics*, Lectures XXXIX, XL; Mansel's *Prolegomena Logica*, ch. v; Mill *on Hamilton*, ch. xvi; Lewes' *History of Philosophy*, Articles on Hume and Kant. I refer only to books likely to be within the student's reach. In quoting or referring to Hume, I have employed only his *Essays*. Many writers persist in making references to his *Treatise of Human Nature*, a work written at the early age of twenty-seven and afterwards repudiated by the author as containing an immature expression of his opinions [29]. In the *Advertisement*

[29] This work is undoubtedly of the highest philosophical interest,

to his Essays, he desires that 'the following Pieces may alone be regarded as containing the author's philosophical sentiments and principles.'

Note 3.—That a cause is . . ..; that every event has a cause; that the same cause is always attended with the same effect; are obviously three distinct propositions, and still there are few writers who, in their treatment of the question of Causation, have not more or less confounded them. The first proposition (if completed) would be the Definition of Cause, the predicate, of course, depending on the view adopted with reference to the question discussed in the previous note. The second is a statement of the Law of Universal Causation, the third of the Law of the Uniformity of Nature.

It will be observed that in the text of this chapter I have said of each of these laws that it 'is universally admitted by mankind, or, at least, by the reflecting portion of mankind.' The latter clause must be regarded as emphatic, and suggests, I think, a sufficient answer to those authors who call in question their universal reception. Mr. Lewes, speaking of the Law of Universal Causation, says, 'All believe irresistibly in particular acts of causation. Few believe in universal causation;

but when we are concerned in determining the matured philosophical opinions of Hume it cannot be regarded as authoritative. It is curious to find a recent editor of Hume's *Essays* expressly defending the practice on which I have animadverted in the text. See Mr. Grose's edition of Hume's *Essays*, vol. i. p. 39.

and those few not till after considerable reflexion[20].' He then proceeds to adduce the case of a student of chemistry, who could not be convinced of the truth of the Law, but 'looked upon the argument as an unwarrantable assumption.' Now I venture to suggest that this incapacity was due to the terms of the proposition not being made sufficiently intelligible to him. I question whether any man of average powers of understanding could be found who would maintain the contradictory of either of these Laws; who would assert, that is to say, that an event might happen without anything to account for it, or that a repetition of exactly the same circumstances might be followed by a different effect. That a considerable amount of intelligence is necessary in order to understand the general terms in which the propositions are stated, is undeniable, but, when once understood, I presume that the propositions cannot fail to be acquiesced in. Like all other propositions, however, of wide import, they may be both understood and acquiesced in, without being fully realised. It is the full and constant realisation of these Laws, at all times and under all circumstances, which mainly distinguishes the man of scientific from the man of unscientific habits of thought. The unscientific man either does not think of enquiring into the causes of the phenomena around him, or notes with little precision the circumstances which he is investigating. The scientific man, on the other hand, insists on invariably referring the phenomena in which

[20] Lewes' *History of Philosophy*, Article on Kant.

he is interested to their several causes, and is satisfied with nothing but the most rigorous enquiry into the relation between these causes and their effects.

But, it may be asked, if the Laws of Universal Causation and of the Uniformity of Nature are, on reflexion, thus universally received, by what mental process do men assure themselves of their truth? Of the origin of these, as of kindred beliefs, two different explanations are offered by rival schools of psychologists. According to one school, the human mind is so constituted that it cannot but accept them; they are fundamental beliefs which exist in the mind prior to all experience, though it is experience which occasions us to realise our possession of them. We have never *learnt* them; we have simply discovered that we possess them. Thus Reid, speaking of our conviction that the future will resemble the past[31] (what we now call the Law of Uniformity of Nature), says, 'The wise Author of our nature hath

[31] This, however, is a very inadequate statement of the Law of the Uniformity of Nature. 'It has been well pointed out,' says Mr. Mill, 'that Time, in its modifications of past, present, and future, has no concern either with the belief itself, or with the grounds of it. We believe that fire will burn to-morrow, because it burned to-day and yesterday; but we believe, on precisely the same grounds, that it burned before we were born, and that it burns this very day in Cochin-China. It is not from the past to the future, *as* past and future, that we infer, but from the known to the unknown; from facts observed to facts unobserved; from what we have perceived, or been directly conscious of, to what has not come within our experience. In this last predicament is the whole region of the future; but also the vastly greater portion of the present and of the past.'—Mill's *Logic*, Bk. III. ch. iii.

implanted in human minds an original principle by which we believe and expect the continuance of the course of nature, and the continuance of those connexions which we have observed in time past. It is by this general principle of our nature, that, when two things have been found connected in time past, the appearance of the one produces the belief of the other[32].' And Dr. Whewell, speaking of the Law of Universal Causation, says, 'We assert that "Every event must have a cause:" and this proposition we know to be true, not only probably, and generally, and as far as we can see; but we cannot suppose it to be false in any single instance. We are as certain of it as of the truths of arithmetic or geometry. We cannot doubt that it must apply to all events past and future, in every part of the universe, just as truly as to those occurrences which we have ourselves observed. *What* causes produce what effects;—what is the cause of any particular event;—what will be the effect of any peculiar process;—these are points on which experience may enlighten us. Observation and experience may be requisite, to enable us to judge respecting such matters. But that every event has *some* cause, Experience cannot prove any more than she can disprove. She can add nothing to the evidence of the truth, however often she may exemplify it. This doctrine, then, cannot have been acquired by her teaching[33].'

---

[32] Reid's *Inquiry into the Human Mind on the Principles of Common Sense,* ch. vi. § 24.

[33] Whewell's *History of Scientific Ideas,* Bk. III. ch. ii. § 1.

The opposite school of psychologists (of which Mr. Mill and Professor Bain may be taken as the modern representatives) maintains that there is nothing in these and kindred beliefs which compels us to distinguish them generically from other truths, but that, like all other truths, they are the result of Experience. From our earliest years, we have been so constantly accustomed to observe one change preceded by another change, and the same antecedents followed by the same consequents, as well as to find our own experience in these respects corroborated by that of others, that, on reflexion, we *all* acquiesce, and *cannot but* acquiesce, in the statements which generalise these facts. This, it is held, is a sufficient explanation of that *universality* and *necessity*, which, by the advocates of the intuitional theory, described in the last paragraph, are supposed to distinguish the 'fundamental beliefs of the human mind' or 'the principles of common sense,' as they are called by these authors, from all other truths. The beliefs have acquired the character of universality and necessity, not because they have sprung from any other source than our ordinary beliefs, but because of the constancy and variety of the experience from which they are gained. 'In fact, our whole lives,' says James Mill, 'are but a series of changes, that is, of antecedents and consequents. The conjunction, therefore, is incessant; and, of course, the union of the ideas perfectly inseparable. We can no more have the idea of an event without having the ideas of its antecedents and its consequents, than we can have the idea and not have it at the same time [34].' But here occurs a difficulty. If the Laws of Universal Causation and of the Uniformity of Nature are inferred from particular facts of causation, are generalisations from experience, or, in other words, inductions, how is it that they are made the grounds of all other inductions? Is not this to argue in a circle? The answer to this difficulty is that the Laws in question are the result of an uniform and constant experience, co-extensive not with the life of the single individual who employs them, but with the entire history of the human race; that, consequently, when we adduce them as the grounds on which our other inductions rest, we are performing the perfectly legitimate process of resolving narrower into wider cases of experience. The argument, in short, is this: the inference from this narrow field of observation (the particular induction which we happen to be making) must be allowed to be true, unless we are prepared to deny one or other of the much wider generalisations which constitute the Laws of Universal Causation and of the Uniformity of Nature. To recur to the instance adopted in the text, the proposition that bodies, subject to the action of gravity only, fall through equal vertical spaces in equal times, can be called in question only on

---

[34] James Mill's *Analysis of the Phenomena of the Human Mind*, ch. xi. The position maintained by James Mill is that these beliefs owe their universality to the fact of their being inseparably associated with all our other cognitions. This is only another mode of stating the theory which derives them from experience.

peril of doubting one or other of the fundamental laws; thus, the doubt which might attach to it is shifted to two other propositions which no one would think of questioning.   Or, to state the same position in a slightly different form, this particular instance is shown to be a member of an infinitely long series, the other members of which have been examined and approved; as, therefore, it differs in no essential respect from them, it claims to be admitted also.   There is, indeed, throughout this argument one assumption; as the rival theory assumed the trustworthiness of what it styled our 'fundamental beliefs,' so this assumes the validity of experience.   But, unless we make one or other of these assumptions, we must be prepared to maintain that knowledge is altogether impossible [35].

There is a third theory of the origin of universal beliefs which combines, with certain modifications, both the others.   It would admit that all beliefs alike are ultimately derived from experience, and still it would freely adopt the language that there are some beliefs which are 'native to the human mind.'   The word 'experience,' as ordinarily employed by psychologists,

[35] It should be noticed that Dr. Mansel, while agreeing in the main, as he usually does, with the intuitional school, in respect to the origin of our belief in the Law of Universal Causation, refers to experience the origin of our belief in the Uniformity of Nature. 'The belief in the uniformity of nature is not a necessary truth, however constantly guaranteed by our actual experience.' Mansel's *Metaphysics*, Chapter on Necessary Truths.   Cp. *Prolegomena Logica*, ch. v.   Dr. Mansel's treatment of these questions is, in many respects, peculiar to himself.

includes not only the experience of the individual, but the recorded experience of mankind. On the theory, however, of which I am now speaking, it has a still more extended meaning; it includes experience, or, to speak more strictly, a peculiar aptitude for forming certain experiences, *transmitted* by hereditary descent from generation to generation. While some ideas occur only to particular individuals at particular times, there are others which, from the frequency and constancy with which they are obtruded upon the minds of men at all times and under all circumstances, become, after an accumulated experience of many generations, *con-natural*, as it were, to the human mind. We assume them, often unconsciously, in our special perceptions, and when the propositions, which embody them, are propounded to us, we find it impossible, on reflexion, to doubt their truth. It is by personal experience of external objects and their relations that each man re-cognises them, but the *tendency* to recognise them is transmitted, like the physical or mental peculiarities of race, from preceding generations, and is anterior to any special experience whatever on the part of the individual. This theory, to which much of modern speculation ap-pears to be converging, is advocated with great ability in the works of Mr. Herbert Spencer[36].

The student who wishes for further information on the questions discussed in this Note is referred to Dugald Stewart's *Philosophy of the Human Mind*, Part II. ch. v.

[36] See especially his work on the *Principles of Psychology*.

§ 2 [37] ('Of that Permanence or Stability in the order of Nature which is presupposed in our Reasonings concerning Contingent Truths'); Reid's *Intellectual Powers*, Essay VI. ch. vi; Reid's *Active Powers*, Essay I. ch. iv; Hamilton's *Supplementary Dissertations to Reid's Works*, Note A, § 3, Note Q; Hamilton's *Lectures on Metaphysics*, Lectures XXXIX, XL; James Mill's *Analysis of the Phenomena of the Human Mind*, ch. xi; Mill's *Logic*, Book III. ch. iii–v, xxi; Mansel's *Prolegomena Logica*, ch. v; Mansel's *Metaphysics*, Section on Necessary Truths; Mill *on Hamilton*, ch. xvi; Lewes' *History of Philosophy*, Article on Kant; Bain's *Moral and Mental Science*, Book II. ch. vi, with Appendix B; Herbert Spencer's *Principles of Psychology*. The student, in employing these references, must be careful to distinguish between what relates to the Law of Universal Causation (sometimes called the Principle of Causality) and the Law of the Uniformity of Nature. The two Laws, as already noticed, are not always distinguished with sufficient care.

[37] In Sir W. Hamilton's edition of Stewart's Works, the corresponding reference is Part II. Subdivision I. ch. ii. section 4, subsection 2.

CHAPTER II.

Of Processes subsidiary to Induction.

OF the various mental processes subsidiary to Induction proper, it will be sufficient for our purpose to discuss Observation and Experiment, Classification (including Nomenclature and Terminology), and Hypothesis.

§ 1. Of Observation and Experiment.

These words are now so familiar, that they hardly require any explanation. To *observe* is to watch with attention phenomena as they occur, to *experiment* (or, to adopt more ordinary language, to *perform an experiment*) is, not only to observe, but also to place the phenomena under peculiarly favourable circumstances, as a preliminary to observation. Thus, every experiment implies an observation, but it also implies something more. In an experiment, I arrange or create the circumstances under which I wish to make my observation. Thus, if two bodies are falling to the ground, and I attend to the phenomenon, I am said to *observe* it, but, if I place the bodies under the exhausted receiver of an air-pump, or cause them to be dropped under any special circumstances whatever, I may be said not only to make an observation, but also to perform an experiment. Bacon has not inaptly compared experiment with the torture of witnesses[1]. Mr. Mill distinguishes between the two processes, by saying that in observation we *find* our instance in nature, in experiment we *make* it, by an artificial arrangement of circumstances. 'When, as in astronomy, we endeavour to ascertain causes by simply watching their effects, we *observe;* when, as in our laboratories, we interfere arbitrarily with the causes or circumstances of a phenomenon, we are said to *experiment*[2].'

As Observation often involves little or no conscious effort, while Experiment always implies an artificial arrangement of circumstances, it might be expected that the general employment of the former for scientific purposes would long precede that of the latter. And this supposition is confirmed by the History of Science. Though it is false to affirm that Experiment was never employed by the Greeks[3], its general neglect was certainly one cause of the little progress made by them in the physical sciences.

In the attempt to ascertain the effect of a given cause, there can be no question of the general superiority of

---

[1] 'Quemadmodum enim in civilibus ingenium cujusque, et occultus animi affectuumque sensus, melius elicitur, cum quis in perturbatione ponitur, quam alias : simili modo, et occulta naturæ magis se produnt per vexationes artium, quam cum cursu suo meant.' *Nov. Org.*, Bk. I. Aph. xcviii.

[2] Thomson and Tait's *Natural Philosophy*, vol. i. § 369.

[3] For a refutation of this popular misconception, see Mr. Lewes' work on *Aristotle*, ch. vi. Mr. Lewes, however, seems to me not sufficiently to recognise the slight extent to which Experiment was employed in ancient as compared with modern times.

Experiment over Observation. To be able to vary the circumstances as we choose, to produce the phenomenon under investigation in the precise degree which is most convenient to us, and as frequently as we wish, to combine it with other phenomena or to isolate it altogether, are such obvious advantages that it is not necessary to insist upon them. Without the aid of artificial experiment, it would have been impossible, for instance, to ascertain the laws of falling bodies. To disprove the old theory that bodies fall in times inversely proportional to their weights, it was necessary to try the experiment; to be able to affirm with certainty that all bodies, if moving in a non-resisting medium, would fall to the earth through equal vertical spaces in equal times, it was essential to possess the means of removing altogether the resisting medium by some such contrivance as that of the air-pump. In some of the sciences, such as Chemistry, the Sciences of Heat, Light, and Electricity, it is next to impossible, at least in their inductive stage, to advance a single step without the aid of Experiment. No amount of mere Observation would ever have enabled us to detect the chemical elements of which various bodies are composed, or to ascertain the effects of these elements in their pure state. Even when Observation alone reveals to us a fact of nature, Experiment is often necessary in order to give precision to our knowledge. That the metals are fusible, and that some are fusible at a lower temperature than others, is a fact which we can conceive to have been obtruded upon man's observation, but the precise temperature at which each metal begins to change the solid for the liquid condition could be learned only by artificial experiment.

But, though, in ascertaining the effect of a given cause, Experiment is a far more potent instrument than Observation, the latter process is also available, and is frequently of the greatest service. Thus, the Science of Medicine equally avails itself, for this purpose, both of observations and experiments. The scientific physician will not only *try* the effects of different medicaments, different modes of diet, and the like, but he will also *watch* the effects on the organic system of various occupations, habits, and pursuits. In some cases even, as in all astronomical and many physiological phenomena, the only means open to us of ascertaining the effect of a given cause is Observation. If we wish to ascertain the various phenomena attendant on a shower of meteors or a total eclipse of the sun, we must wait till the shower of meteors occurs or the total eclipse takes place. If we wish to learn the effects of the lesion of a particular part of the nervous system, we must generally wait till an instance offers itself; there are many experiments too dangerous and too costly to be made, at least in the case of man.

While, however, both Observation and Experiment are available in ascertaining the effects of a given cause, in the reverse process of ascertaining the cause of a given effect, Observation alone is open to us. 'We can take a cause,' says Mr. Mill, 'and try what it will produce; but we cannot take an effect, and try' [that is, experimentally] 'what it will be produced by. We can only watch till we see it produced, or are enabled to produce it by accident.' In those cases, consequently, in which effects alone are patent to us, and the causes are concealed from our view, we are compelled, unless we are able to reverse the problem in the manner noticed in the next paragraph, to have recourse to Observation. A new disease makes its appearance: the mode of its action, and the conditions favourable or unfavourable to its diffusion, can only be learned by a careful observation and comparison of cases.

It should, however, be noticed that the problem of finding the cause of a given effect is, in practice, as, for instance, in many cases of chemical analysis, often reversed, and that, by setting in action a variety of causes, we try to discover whether any one of them will produce the effect in question. Experiment is thus substituted for Observation.

It will readily be seen that those Sciences which depend wholly or mainly on Observation are, as inductive sciences, at a great disadvantage compared with those in which it is possible largely to employ Experiment. Where we wish to ascertain the effect of a given cause, and we cannot make the instances for ourselves, the want of appropriate and definite instances will often completely baffle us. And, though the cause of a given effect can only be learned by Observation, this is generally an enquiry of extreme difficulty, requiring to be supplemented by experiment, or the actual production of the given effect by the supposed cause, before we can be certain that it has been conducted with the required accuracy. Thus, mere observation of the electrical phenomena which we witness in the heavens could never have given us the Science of Electricity. The experiments which we may conduct in an hour are often worth a century spent in observations.

In the Science of Astronomy this defect is more than compensated by the extreme simplicity of the phenomena, the heavenly bodies being regarded by us, not in themselves, but only in their mutual relations. Hence, we are, at a comparatively early stage, enabled to apply the Deductive Method, and to solve the problems of Astronomy by mathematical calculations. But in the very complex Science of Physiology this resource is not open to us, and hence the backwardness of those departments of physiological science in which direct experiment is not available. Any animal or vegetable organism is so complex, the data are so numerous, and bear to each other so many different relations, that, hitherto, it has been found impracticable to subject physiology, at least in any detail, to a deductive treatment. In social and political speculations, the want of experiment is, to some extent, supplied by statistics. A social or political experiment is generally as impracticable as an experiment in physiology, and the danger with which it is attended is often incomparably greater. But the number of observations open to us in these enquiries (as, for instance, in respect to crime, education, trade, taxation, &c.) is often very large, and, by carefully comparing and systematising them, we may frequently detect some relation between two circumstances which enables us, with great probability, to infer that one has something to do with the production of the other. I am here, however, trenching on the province of those chapters which treat more peculiarly of inductive inference.

The following Rules may be laid down for the right conduct of Observations and Experiments :—

Rule I. They must be *precise*. It is often of the utmost importance to notice the exact time at which an event occurs, the length of its duration, the position of an object in space, its relation to surrounding objects, and the like. We are all acquainted with the prime importance of precision of detail in legal evidence; it is no less indispensable in scientific research. For the purpose of enabling us to attain this object, various instruments and methods have been invented. As instances of these devices may be given, amongst instruments, the telescope, the microscope, the thermometer, the barometer, measures of various kinds, the balance, the dial, the clock, the watch, the chronometer, the vernier, the goniometer, the galvanometer, the thermo-electric pile; amongst methods, the decimal system of notation, fractions both vulgar and decimal, the divisions of time, the various contrivances for the measurement of space, the method of double-weighing, the method of least squares, the personal equation in astronomical observations. To these instances might be added numerous others, but these will be sufficient to show the great aid derived by what may be called the natural methods of observation from artificial contrivances. The Thermometer and the Method of Double-Weighing furnish such striking exemplifications of the assistance thus afforded, that, though they are probably familiar to most of my readers, it may be desirable to explain them, one as an example of an instrument, the other of a method.

The Thermometer (it is not necessary here to describe the different kinds of thermometers) is a contrivance for determining the degree of temperature, irrespective of the mode in which it affects individual organisms. As our sensibility varies considerably under different circumstances, so that what at one time affects us with the sensation of hot will at another affect us with that of cold, the sense of touch cannot be depended upon for giving us accurate measurements of temperature. But the fact that, while the pressure remains unaltered, an augmentation of temperature, with certain rare exceptions (to be noticed hereafter), expands the bodies subject to its influence furnishes us with such a means of measurement. We take a substance which notably exemplifies the power of heat in expansion, such as mercury, alcohol, or, where it is necessary to ensure great precision, atmospheric air carefully prepared, and, by confining it within a tube, and marking off a scale of measurements along the side, we are enabled, by noting the degree of expansion of the substance in the tube, to estimate, at least approximately, the degree of temperature in the atmosphere or any other body, the conditions of which we are investigating.

The method of Double-Weighing is peculiarly simple and ingenious. It is a contrivance for remedying any possible inequality in the effective arms of the beam of the Balance. The body to be weighed is placed at one end of a balance, and is exactly balanced by another body placed at the other end; the first body is then removed, and its place supplied by a standard weight or weights, till these exactly balance the second body; we are thus, on the principle that things which are equal to the same thing are equal to one another, assured of the precise equivalence in weight of the body to be weighed and the standard weight or weights, provided, of course, that we can depend on the instrument giving the same results in successive weighings.

It frequently happens, however, that a single observation may greatly mislead us. I may be in a district at one time, and find the air very temperate and agreeable; the next time I come, it may be peculiarly hot, or chill, or moist. I may see a man, at the first shot, hit his mark; but, at the subsequent shots, he may fire very wide of it. Hence the importance, whenever there is any liability to error, of taking an *average of observations.* If a sufficient number of observations be taken, there is every probability that an error in one direction will be compensated by an error in the other, and that an average, derived from all the observations, will approximate much more nearly to the truth than any single observation is likely to do. Thus, if I wish to ascertain the true character of the climate at any particular place, the observations I consult must extend over a considerable number of years; if I wish to estimate truly the skill of the marksman, I must watch, not a single shot, but many successive ones. The average, it is true, is liable to error, but any single observation is much more so. There is hardly any department of science, depending upon observation, in which, if it be our object to obtain precision, this method is not indispensable [4].

 Rule II. But, though it is necessary to be precise in our observations and experiments, it is also important, in order to avoid distraction and waste of time, to attend only to the *material circumstances* of the case we are investigating. A physician, for instance, in prescribing for his patient, would not now think it necessary to take an observation of the planets, nor would a chemist, in gathering herbs for his decoctions, think it of any consequence to notice the phase of the moon. A caution should, however, be added. Before neglecting any circumstance in our observations, it is of the utmost im-

---

[4] The student who may wish for further information in connexion with Rule I. is referred to Dr. Whewell's *Novum Organon Renovatum*, Bk. III. ch. ii., and Herschel's *Discourse on the Study of Natural Philosophy*, § 387–9. On the importance of taking an average of observations, see Herschel's *Discourse*, § 226–30.

portance to have ascertained beyond doubt that it is *not* material to the subject of our enquiries[5]. To neglect this caution would be a violation of the first Rule.

Rule III. The *circumstances* under which an observation or experiment is made should, except in the very simplest cases, be *varied* as much as possible. A physician, in studying the character of a disease, will note its effects on persons of different ages, constitutions, habits of life, and the like. An astronomical observer will not be content with a single observation of a newly-discovered comet, but will note the phenomena which attend it at various stages in its passage through the heavens. A chemist will combine a newly-discovered element with the various other elements, and will try upon it the effect of heat, pressure, &c. It is, of course, implied that some discretion will be employed in the application of this Rule, and that the variation of circumstances will not be carried beyond the point at which there is some probability of its adding to our knowledge.

[5] The neophyte in science requires to be reminded that observations which might at first be supposed to be *immaterial* are often afterwards found to be amongst the circumstances most *material* to the enquiry. 'Could anything' (says Dr. Rolleston, in his Address before the Medical Association in 1868) 'have seemed at first sight to be more impertinent, more otiosely curious and trifling, than to enquire during an epidemic of cholera what was the nature of the subsoil in the area it was ravaging, to what depth it was porous, and at what level the water was, and had been previously, standing in it? Yet, as I think, Von Pettenkofer has at last fought out and won his battle on these points.'

Rule IV. The phenomenon under investigation should if possible be *isolated* from all other phenomena, or, at least, from all those which are likely to interfere with our study of it. In studying the effects of the action of gravity upon bodies, it was necessary to exhaust the atmosphere and to withdraw the support, and, by thus *isolating* the phenomenon, to enable us to perceive how bodies behave when subject to the action of gravity only. A physician, in trying the effects of a new drug, will, at first at least, administer it alone, and not in combination with other drugs which might augment or counteract its influence on the system[6].

A beautiful instance of the isolation of a phenomenon is afforded whenever there occurs a total eclipse of the sun. As, on these occasions, the moon, by a curious coincidence, exactly covers, or rather more than covers, the sun's surface, and thus intercepts all light from it, we are able to see certain rose-coloured protuberances, projecting, as it were, from the dark edge of the moon, but, in fact, proceeding from the sun. The real nature of these 'red flames' was long a matter of dispute, but it seems now to be conclusively settled that they are portions of an atmosphere of incandescent hydrogen in which the sun is enveloped, and which often shoots out

---

[6] By observing the third of these rules we usually prepare our instances for the application of what will hereafter be explained as the Method of Agreement, and by observing the fourth for the application of what will hereafter be explained as the Method of Difference.

in these flames to distances estimated, on one or two occasions, at no less than 300,000 miles[7]. They can now be seen, whenever the sun is shining, by means of the spectroscope; but had it not been for the isolation of the phenomenon thus produced by the intervention of the moon, astronomers would have till quite recently been ignorant of its existence. Here, to use a bold metaphor, we might say that Nature herself performs an experiment for us.

When it is impossible entirely to isolate a phenomenon, it is sometimes possible so far to diminish the action of 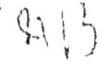 the concomitant circumstances as to be able accurately or approximately to calculate what would be the effect, if they were altogether absent. Thus, we can never altogether remove the influence of friction on a moving body, but we can so far diminish it as to be able to say what the effect would be were no such influence at work. We cannot altogether eliminate the influence of extraneous circumstances on a patient subject to medical régime, but, by due care, we may minimise the excitement, fatigue, ennui, or other unfavourable conditions which might interfere with our treatment.

The circumstances under which we perform our experiments being more in our own power than those under which we conduct our observations, it is obvious that the foregoing rules, and especially the third and fourth, can be more easily observed in experiments than in observations.

[7] See Young on the Sun, p. 202.

§ 2. *On Classification, Nomenclature, and Terminology.*

(1) OF CLASSIFICATION.

A classification, in the widest sense of the term, is a division, or a series of divisions and subdivisions[8]. The process of classifying our own thoughts or feelings, or the actions of ourselves or others, or the external objects which surround us, is one of the most constant occupations of the mind. Thus, we are perpetually dividing outward objects into those which are useful or those which are useless or noxious to us; those which are useful into such as are within and such as are beyond our power to attain; those which are useful and which it is within our power to attain into such as are to be sought at once, and those the effort to appropriate which may be more advantageously postponed,—each of these divisions admitting of almost infinite subdivision. In fact, as has frequently been remarked, every attribution of a general name implies an act of division or classification. When we speak of a horse, we are dividing all objects into those which are horses and those which are not. When we speak of a bay horse, we are superadding to this division the subdivision of horses into bay horses and those of any other colour.

But the process of Classification of which I am about to treat, though the same in kind with that which we employ in the affairs of ordinary life, is of a much more complex and systematic character. The great difference is that, whereas in the affairs of ordinary life we generally

[8] See *Deductive Logic*, Part II. ch. viii.

classify objects with reference to some one principle, that principle varying according to the particular purpose we happen to have in view (thus we classify horses according to their colour, their breed, their strength, &c., each classification being suggested by some distinct purpose), a scientific classification must take account of all the points of difference which are in any way likely to facilitate the scientific investigation of the group. The purport of the science being defined, the classification must be based, not on one or two characters, selected arbitrarily, but on the entire assemblage of characters which the science investigates. Thus, if Botany be defined as the science which investigates the organisation (including under that term the form, structure, and functions) of plants, a botanical classification, in order to be strictly scientific, must not omit to take into account any part of that organisation. But it is evident that such a requirement would produce endless confusion, unless we could discover some mode of subordinating the characters, so as to make the more important points of difference the basis of the higher divisions in the series. Hence we see already that a scientific classification must be guided by at least two principles, a review of all the characters or distinguishing marks, so far as they are known and so far as they fall within the scope of the science, and a subordination of these characters one to another. To these principles others will subsequently be added.

Before proceeding to the attempt to ascertain inductively facts of co-existence or causation amongst a vast mass of phenomena, it is often highly important, if not essential, to arrange these phenomena in groups, as well as to determine the order in which these groups themselves shall be arranged. Hence the importance of laying down correct rules for Classification in a System of Inductive Logic. It is exclusively as subsidiary to Induction that I shall here consider the subject of Classification[9].

A scientific Classification, regarded as subsidiary to Induction employed for scientific purposes, may be defined as *A Series of Divisions, so arranged as best to facilitate the complete and separate study of the several groups which are the result of the divisions, as well as of*

[9] It will probably occur to the student that the materials for Classification can themselves only be obtained by Induction. And this is true. All Classification implies the prior employment of an *Inductio per Enumerationem Simplicem,* by which we establish the fact of the co-inherence of certain attributes. But these 'inductions of co-existence' (see pp. 7–9), which precede our classifications, are altogether of a different order from the 'inductions of causation' which it is the ultimate aim of science to establish, and to which I regard Classification as mainly subordinate. I say 'mainly subordinate,' for, of course, there is no doubt that, when, by means of certain 'inductions of co-existence,' we have constituted a class, we are in a more favourable position than before for detecting additional facts of co-existence among the associated phenomena.

When, from a wide experience, I find that the attributes a, b, c, d, e invariably co-exist in the same objects, I generally constitute these objects into a class, and designate them by a class-name. The name thus serves to recall the fact of the co-inherence of the attributes, and I am far more likely, than if I had never made the classification, to discover the co-existence with the other five of some sixth attribute, say f, or to be able to trace some causal connexion between, say, a and b, or a, b, and c.

the entire subject under investigation. 'The general problem of classification,' says Mr. Mill[10], 'in reference to these [namely, scientific] purposes may be stated as follows: To provide that things shall be thought of in such groups, and those groups in such an order, as will best conduce to the remembrance and to the ascertainment of their laws.'

The sciences of Botany and Zoology are rightly regarded as furnishing the best examples of Scientific Classification. The excellence of the classifications which they present may be referred to two reasons. The first is the extraordinary multiplicity of the different kinds of animals and plants which are found on the surface of the globe: this fact has, from the earliest times, exercised man's ingenuity in the attempt to name them and reduce them to order. The second reason may be found in the imperfection of these sciences in their present condition: the difficulty of discovering laws of succession, or, in other words, relations of cause and effect, in the animal and vegetable kingdoms has naturally led scientific enquirers to concentrate their attention on the far easier task of describing and arranging the objects themselves. Mineralogy, though its classifications are less systematic and complete, is also, in the present state of the science, mainly occupied in attempting the work of classification.

The best means, perhaps, of making the student acquainted with the nature of scientific classification is to compare the method of *natural classification* (which aims

---

[10] Mill's *Logic*, Bk. IV. ch. vii. § 1.

at being strictly scientific) with that of *artificial classifica-tion* (which, so far as it is artificial, is not scientific), giving illustrations from the sciences of Botany and Zoology. An examination of the natural system will enable us to lay down certain rules for scientific classi-fication, and I shall conclude with such remarks as may seem necessary in order to preserve the student from erroneous impressions.

(A natural system of Classification aims at classifying objects according to the whole of their resemblances and differences, so far as these are recognised by the science in whose service the classification is made.) But amongst these resemblances and differences some are found to be invariably attended by a number of others, and conse-quently these, as the *more important*, are selected as the characters by which to discriminate the higher divisions of the series, the less important characters being, through-out the whole series, subordinated to the more important. This successive subordination of characters and the con-sequent coincidence of the groups formed by our classi-fications with what appear to be the great divisions of nature are the peculiarities which mainly distinguish a natural system. (An artificial system, on the other hand, is one which selects arbitrarily some point of difference amongst the objects to be classified, and then, so far as possible, makes this or similar points the basis of its classifications.) No system, however, as we shall see presently, is purely artificial. Though of little use, except as a preliminary effort, for the purposes of science, an artificial system possesses one great advantage. (As it bases its divisions, where possible, on some one property, and that generally something which at once strikes the eye (one of the earliest of the modern attempts to classify plants took for its basis the form of the corolla), it is peculiarly easy of application, and can be much more readily learnt than a natural system.) It thus often serves the purposes of a key, by which we may easily discover the place of a group in a natural system. I now proceed to offer illustrations.

In Botany, the most celebrated artificial system is that known as the Linnæan, though Linnæus also did much towards the establishment of a natural system. In this system, which was a great advance on preceding artificial systems, the main basis of classification is the number of stamens and pistils which are to be found in the flowering plant. This character is, however, to some extent modified by other considerations, such as the relative lengths of the stamens, the shape of the fruit, &c.; so far as these modifications are admitted, the Linnæan system approaches to a natural system. The annexed Tables (extracted from Balfour's *Manual of Botany*[11]) will give the student some idea of the manner in which the Classes (higher divisions) and the Orders (divisions intermediate between the Classes and Genera) are constituted according to the Linnæan system. It should be premised that the stamens are the male organs, and the pistils the female organs of a plant.

[11] §§ 716, 717.

Tabular View of the Classes of the Linnæan System.

A. Flowers present, or evident Stamens and Pistils (Phanerogamia).

  I. Stamens and Pistil in every flower.

    1. Stamens free.

      a. Stamens of equal length, or not differing in certain proportions ;

in number 1	Class I.	Monandria.
— 2	II.	Diandria.
— 3	III.	Triandria.
— 4	IV.	Tetrandria.
— 5	V.	Pentandria.
— 6	VI.	Hexandria.
— 7	VII.	Heptandria.
— 8	VIII.	Octandria.
— 9	IX.	Enneandria.
— 10	X.	Decandria.
—12–19	XI.	Dodecandria.
— 20 } inserted on Calyx—	XII.	Icosandria.
or more } on Receptacle...	XIII.	Polyandria.

      b. Stamens of different lengths ;

two long and two short	XIV.	Didynamia.
four long and two short	XV.	Tetradynamia.

    2. Stamens united ;

by Filaments in one bundle...	XVI.	Monadelphia.
—— in two bundles	XVII.	Diadelphia.
—— in more than two bundles	XVIII.	Polyadelphia.
by Anthers (Compound flowers)	XIX.	Syngenesia.
with Pistil on a Column	XX.	Gynandria.

  II. Stamens and Pistil in different flowers ;

on the same Plant	XXI.	Monœcia.
on different Plants	XXII.	Diœcia.

  III. Stamens and Pistil in the same or in different flowers on the same or on different Plants } XXIII. Polygamia.

B. Flowers absent, or Stamens and Pistils not evident } XXIV. Cryptogamia.

The Classes are sub-divided into Orders, as will be seen from the next Table, on a less uniform plan than that on which they were themselves constituted.

TABULAR VIEW OF THE ORDERS OF THE LINNÆAN SYSTEM.

Class I.	Monogynia[12]	1	Free Style.
II.	Digynia	2	Free Styles.
III.	Trigynia	3	—
IV.	Tetragynia	4	—
V.	Pentagynia	5	—
VI.	Hexagynia	6	—
VII.	Heptagynia	7	—
VIII.	Octogynia	8	—
IX.	Enneagynia	9	—
X.	Decagynia	10	—
XI.	Dodecagynia	12–19	—
XII.	Polygynia	20 and upwards.	
XIII.			

XIV. Gymnospermia .... Fruit formed by four Achænia.
Angiospermia .... Fruit, a two-celled Capsule with many seeds.

XV. Siliculosa .... Fruit, a Silicula.
Siliquosa .... Fruit, a Siliqua.

XVI.
XVII. Triandria, Decandria, &c. (number of Stamens), as in Classes.
XVIII.

XIX.
Polygamia Æqualis ...... Florets all hermaphrodite.
———— Superflua ...... Florets of the disk hermaphrodite, those of the ray pistilliferous and fertile.
———— Frustranea ... Florets of the disk hermaphrodite, those of the ray neuter.
———— Necessaria ... Florets of the disk staminiferous, those of the ray pistilliferous.
———— Segregata...... Each floret having a separate involucre.
Monogamia .... Anthers united, flowers compound.

[12] It must not be supposed that all the Orders, Monogynia, &c., exist in each of the first thirteen Classes. When an Order is absent, the next Order which is present takes its place in the numerical arrangement. Thus, if the Order Trigynia be absent, and the next Order which is present be Tetragynia, as in Class IV, this latter will rank as the third Order.

XX.		
XXI.	Monandria, Diandria, &c.	(number of Stamens), as in the Classes.
XXII.		
	Monœcia	Hermaphrodite, staminiferous and pistilliferous flowers on the same plant.
XXIII.	Diœcia	on two plants.
	Triœcia	on three plants.
	Filices.........................	Ferns.
	Musci ,......................	Mosses.
XXIV.	Hepaticæ	Liverworts.
	Lichenes....................	Lichens.
	Algæ	Sea-weeds.
	Fungi	Mushrooms.

'Even as an artificial method,' says Professor Balfour[13], 'this system has many imperfections. If plants are not in full flower, with all the stamens and styles perfect, it is impossible to determine their class and order. In many instances, the different flowers on the same plant vary as regards the number of the stamens. Again, if carried out rigidly, it would separate in many instances the species of the same genus; but, as Linnæus did not wish to break up his genera, which were founded on natural affinities, he adopted an artifice by which he kept all the species of a genus together. Thus, if in a genus nearly all the species had both stamens and pistils in every flower, while one or two were monœcious or diœcious, he put the name of the latter in italics, in the classes and orders to which they belonged according to his method, and referred the student to the proper genus for the description.'

The species of the Linnæan system coincide with those of the natural system. The same is mostly the case with

[13] Balfour's *Manual of Botany*, § 718.

the genera, or next higher divisions. The Linnæan system is, therefore, far from being purely artificial. In fact, when we come to the lower groups of vegetables (genera and species), we are compelled to discriminate them one from another by a multiplicity of characters, so that a purely artificial system of botany would be impossible.

The framers of natural systems of botany, instead of selecting some one character, such as the number of stamens and pistils, as the basis of the higher divisions, attempt to discover a number of characters, any one of which, if employed as the instrument of division, would give the same results as any of the others. This coincidence of divisions founded on various characters is a proof that we have reached some real distinction in nature. The main division of plants into cellular and vascular, or acotyledonous and cotyledonous, and the sub-division of vascular or cotyledonous plants into monocotyledonous and dicotyledonous, furnish remarkable instances of such a coincidence, and may consequently be regarded as corresponding with grand divisions in nature itself.

'In taking a survey of the Vegetable Kingdom, some plants are found to be composed of cells only, and are called *Cellular;* while others consist of cells and vessels, especially spiral vessels, and are denominated *Vascular.* If the embryo is examined, it is found in some cases to have cotyledons or seed-lobes, in other cases to want them; and thus some plants are *cotyledonous,* others *acotyledonous;* the former being divisible into *monocotyle-donous*, having one cotyledon, and *dicotyledonous*, having two [or more] cotyledons. The radicle, or young root of acotyledons, is *heterorhizal*, that of monocotyledons is *endorhizal*, that of dicotyledons, *exorhizal*. When the stems are taken into consideration, it is seen that marked differences occur here also, acotyledons being *acrogenous*, monocotyledons *endogenous*, and dicotyledons *exogenous*. The venation of leaves, parallel, reticulated, or forked, establishes the same great natural divisions; and similar results are obtained from a consideration of the flowers, monocotyledons and dicotyledons being *phanerogamous* and acotyledons *cryptogamous.'*

'Thus, the following grand natural divisions are ar-rived at :—

1. Cellular...Acotyledonous.	Heterorhizal.	Acrogenous.	Crypto-gamous.	
2. Vascular.	Monocotyledonous. Endorhizal.	Endogenous.	Phanero-	
	Dicotyledonous. Exorhizal.	Exogenous.	gamous[14].'	

Having established these Primary Divisions of the vegetable kingdom, the botanist, guiding himself as far as possible by the same principles as those on which the primary divisions were formed, proceeds to divide and sub-divide till at last he arrives at *species,* which are usually defined to be collections of individuals so nearly resembling each other that they may be supposed to be descended from a common stock. Thus, the Class 'Dicotyledones or Exogenæ' is sub-divided into four

[14] Balfour's *Manual of Botany,* §§ 723, 724.

sub-classes, one of which is the 'Thalamifloræ,' characterised as having 'calyx and corolla present, petals distinct and inserted into the thalamus or receptacle, stamens hypogynous.' This sub-class is divided into a number of orders (sixty in Professor Balfour's *Manual*), one of which is Hypericaceæ, the Tutsan or St. John's-wort family, thus described :—

'Sepals 4–5, separate or united, persistent, usually with glandular dots, unequal; æstivation imbricated. Petals 4–5, oblique, often with black dots, æstivation contorted. Stamens hypogynous, indefinite in number; generally polyadelphous, very rarely 10, and monadelphous or distinct; filaments filiform: anthers bilocular, with longitudinal dehiscence; carpels 2–5, united round a central or basal placenta; styles the same number as the carpels, usually separate; stigmas capitate or simple. Fruit either fleshy or capsular, multilocular, and multivalvular, rarely unilocular. Seeds usually indefinite in number, minute, anatropal, usually exalbuminous; embryo usually straight.—Herbaceous plants, shrubs, or trees, with exstipulate entire leaves, which are usually opposite and dotted. Flowers often yellow.'

In this order there are fifteen known genera, one of which is the Hypericum, which is thus described in Irvine's *Handbook of British Plants* :—

'Hypericum, St. John's-wort. Herbaceous plants or shrubs, with opposite simple, entire leaves, which are usually furnished with pellucid dots (reservoirs of essential oil). Sepals five, free or united at the base, ovate, slightly unequal, permanent. Petals as many as the sepals, obtuse, spreading. Stamens indefinite, combined at the base into three or five sets, with small roundish anthers. Ovary with three-five cells or carpels and as many styles, with simple stigmas. Fruit capsular, rarely baccate, three-five-celled, with numerous seeds.'

This genus is divided into sub-genera or sections, one of which is thus described :—

'Stems herbaceous. Stamens in three parcels (triadelphous). Styles three. Capsule three-celled, three-valved.'

The sub-genus or section is again divided into sub-sections, one of which is characterised as having 'stems round, sepals with ciliary glands.' This sub-section contains amongst its species the well-known Hypericum Pulchrum, 'Elegant St. John's-wort,' thus described :—

'Stems erect, bent at the base, round, glabrous, simple or branching. Leaves *ovate, clasping, coriaceous,* smooth, with numerous translucent dots. Flowers in opposite panicled cymes. Sepals *obovate, roundish, with a point, ciliated, with nearly sessile glands.* Petals oblong, ribbed, with black sessile glands [15].'

The first peculiarity which strikes us in these descriptions is the large number of characters which is employed in constituting even the higher divisions of the series. Instead of describing merely the number and distribution of the stamens, as in the Linnæan system, we have, even in the description of the Order, a reference to almost every part of the plant. We next notice the much greater definiteness which the characters assume, as we descend lower in the series. Thus, to take the sepals as an instance, the description of the sub-class simply informs us of the presence of a calyx, while each successive division (except the sub-genus) gives us more and more definite information as to the number, position, form, &c. of the sepals which constitute the calyx.

[15] See Irvine's *Handbook of British Plants,* under Order CIII.

Again, we observe that, in the lower divisions, the stem, leaves, sepals, and petals are the characters which are brought into greatest prominence, whereas the stamens and the various parts of the pistil (the carpels, styles, and stigmas), which are employed in the higher divisions, disappear from the lower, as no longer affording grounds of difference. Now the stamens and pistil, inasmuch as any peculiarity in them is generally accompanied by a larger number of peculiarities in other parts of the plant, are usually of far more importance than the corolla (petals) and calyx (sepals), and therefore it is reasonable to suppose that the grounds of difference furnished by them would be likely to be exhausted in the higher divisions. At the same time we see that, in the instance we have taken, the sepals and petals furnish grounds of difference at a very early stage of the classification, and consequently that even the less important characters are often used concurrently with others to determine the higher divisions.

In Zoology, the advantage of a natural over an artificial classification is more readily recognised than in Botany, the structure and functions of animals being more familiar and apparent than those of plants. A division of animals, for instance, which adopted the number of limbs as its sole distinguishing character, and thus brought together, as *quadrupeds,* the ox and the frog, would be so absurd on the face of it, as to be rejected at once. 'No arrangement of animals,' says Dr. Whewell[16], 'which, in

[16] *History of the Inductive Sciences,* Bk. XVI. ch. vii.

a large number of instances, violated strong and clear natural affinities, would be tolerated because it answered the purpose of enabling us easily to find the name and place of the animal in the artificial system.   Every system of Zoological arrangement may be supposed to aspire to be a natural system.'   He then proceeds to give an instance of an attempt to constitute an artificial classification in the ichthyological branch of Zoology.   'Bloch, whose ichthyological labours have been mentioned, followed in his great work the method of Linnæus,' (who devoted much of his attention to the classification of animals as well as of plants).   'But towards the end of his life he had prepared a general system, founded upon one single numerical principle—the number of fins; just as the sexual system of Linnæus is founded upon the number of stamina: and he made his sub-divisions according to the position of the ventral and pectoral fins; the same character which Linnæus had employed for his primary division.   He could not have done better, says Cuvier, if his object had been to turn into ridicule all artificial methods, and to show to what absurd combinations they may lead.'

'By the *natural method,*' says M. Milne Edwards[17] (whose remarks on Zoological Classifications and the Primary Divisions and Classes of the Animal Kingdom

---

[17] See Milne Edwards' *Zoologie* (in the *Cours élémentaire a'histoire naturelle*), septième édition, §§ 364, 365.   There is an English translation of this work by Dr. R. Knox.   I have followed it, except in a few places where it does not accurately represent the original.

are well worthy of the attention of all students of inductive logic), 'the divisions and subdivisions of the animal kingdom are founded on the whole of the characters furnished by each animal, arranged according to their degree of respective importance; thus, in knowing the place which the animal occupies, we also know the more remarkable traits of its organisation, and the manner in which its principal functions are exercised.

'The rules to be observed in arriving at a natural classification of the animal kingdom are of extreme simplicity, but often there is much difficulty in the application. They may be reduced to two, for the object of the zoologist in establishing such a classification is,—

'1st. To arrange animals in natural series, according to the degree of their respective affinities,—that is to say, to distribute them in such a manner that those species which most nearly resemble each other may occupy the nearest places, while the distance of two species from each other may, in some sort, be the measure of their non-resemblance.

'2nd. To divide and subdivide this series according to the principle of subordination of characters,—that is to say, by reason of the importance of the differences which these animals present amongst them.'

The Primary Divisions of the animal kingdom, according to the natural system, are four, there being four types of structure and of nervous organisation, to which animal life conforms.

'These four principal forms may be understood by a

reference to four well-known animals—the dog, the cray-fish or lobster, the snail, the asterias or sea-star.

'In order that the zoological classification might be a faithful representation of the more or less important modifications introduced into the structure of animals, it was necessary to distribute these beings into four principal groups or divisions; and this is, in fact, what Cuvier did.

'The animal kingdom is divided into *vertebrate animals*, *articulated* or *annulated animals*, *molluscs*, and *zoophytes*.

'The fundamental differences distinguishing these four primary divisions depend chiefly on the mode of arrange-ment of the different parts of the body and on the con-formation of the nervous system. It is easy to under-stand the importance of these two dominant characters: to feel and to move is the especial character of animal life, and these two functions belong to the nervous system. It might readily, then, be anticipated that the mode of conformation of this system would exert a powerful influence over the nature of animals, and would furnish characters of primary importance in classification.

'The general disposition or mode of reunion of the different parts of the body exercises an equally important influence, as modifying the localisation of the functions and the division of the physiological result[18].'

Vertebrate animals are thus described:—

'The *vertebrate animals* resemble man in the more important points of their structure; almost all the parts of their bodies are

[18] Milne Edwards, §§ 372, 373.

in pairs, and disposed symmetrically on the two sides of a medial longitudinal plane; their nervous system is highly developed, and is composed of nerves and ganglions, and of a brain and spinal marrow. To these characteristics we may add that the principal muscles are attached to an internal skeleton, composed of separate pieces, connected together, and disposed so as to protect the more important organs, and to form the passive instruments of loco-motion; that the more important part of this skeleton forms a sheath for the brain and spinal marrow, and results from the reunion of annular portions, called vertebræ; that the apparatus for the circulation is very complete, and that the heart offers at least two distinct reservoirs; that the blood is red; that the limbs are almost always four in number, and never more; finally, that there exist distinct organs lodged in the head for sight, hearing, smell, and taste [19].'

The Primary Division (embranchement) 'Vertebrate Animals' is sub-divided into the five classes, Mammals, Birds, Reptiles, Batrachia, Fishes, of which Mammals are thus described :—

'Organs of lactation. Hot blood. Circulation complete, and heart with four cavities. Pulmonary respiration simple. Lobes of the cerebellum reunited by an annular protuberance. Lower jaw articulated directly with the cranium. The body generally covered with hairs. Viviparous.'

'There exist considerable differences amongst the mammalia, and these modifications of structure serve as the basis for the division of the class into groups of an inferior rank, called *orders*. Most of these groups are so distinct as to admit of no doubt in respect of their limits: they constitute, in fact, natural divisions; but in others the line of demarcation is by no means so distinct.

[19] Milne Edwards, § 374.

Thus a mammal may have points of resemblance so close to two groups as to render it almost indifferent to which it be referred. To some naturalists, differences appear important which are disregarded by others, and hence a want of agreement on the subject of classification has always prevailed.

'The method followed here is nearly the same as that proposed by Cuvier. It rests mainly on the differences mammals show in respect of their extremities and teeth, differences which always imply a crowd of others in habits, structure, and even intelligence.

'Keeping in view the *ensemble* of these characters, the class mammalia may be divided into two groups,—the *monodelphic* and *didelphic.*

'The monodelphic or monodelphian are the more numerous, and are distinguished chiefly by their mode of development. At birth they are already provided with all their organs, and before birth they derive their nourishment from the mother by means of a *placenta*. Their brain is more perfect than the didelphian, by the presence of a corpus callosum uniting the two cerebral hemispheres. Finally, the walls of the abdomen have no osseous supports attached to the margins of the pelvis, as we find in the second great class of mammals. The mammals thus organised have been subdivided into two groups,—namely, *ordinary mammals* and *pisciform mammals*.

'The ordinary mammals are organised principally to live on solid ground; the skin is provided with hairs.

These animals are further subdivided into ten orders: the bimana, quadrumana, cheiroptera, insectivora, rodentia, edentata, carnivora, amphibia, pachydermata, and ruminantia. The first eight of these orders have flexible fingers and toes, with nails covering only the dorsal aspect of the toe or finger, and comparatively small; hence they have been called *unguiculata;* the last two,—namely the pachydermata and ruminantia, have the extremity of the finger and toe entirely enclosed in a hoof; they are thus called *ungulata.*

'The order bimana includes only man: in him alone the arms are destined for prehension, the limbs for progression and support in the erect attitude. Thus, his natural position on the soil is unmistakeably vertical. The teeth are of three kinds, and have their edges on the same plane; they are frugivorous: finally, the brain is more perfect, more highly developed, than in any other animal[20].'

Here the Order is co-extensive with the Species, but usually the Order is divided into Genera, and each Genus into Species. Thus, the Order 'Carnivora' is divided into the Genera 'cat,' 'hyæna,' 'dog,' 'bear,' &c. Again, the Genus 'dog' comprises the dog properly so called, the wolf, and the fox. The Genus 'cat' comprises not only the cat properly so called, but the tiger, lion, panther, &c.

It may be as well to add an account of the characters which distinguish respectively the Order 'Carnivora,' the

[20] Milne Edwards, §§ 409–412.

Genus 'Felis,' and the Species 'Leo,' in order to serve as an example or illustration of the manner in which these several degrees in the scale of Classification are usually described :—

'The order of carnivora is composed of ordinary unguiculated mammals; the form of their dentition is complete, but they have no opposing thumb.   According 'to the mode of life of these animals, their intestinal canal is short; their jaws and their muscles strong, in order to seize and devour their prey; their head from this circumstance seems large.   The jaws are short, thus favouring their strength, and the form of the temporal-maxillary articulation proves that the teeth are made for tearing and cutting, not for grinding or masticating.   The canine teeth are large, long, and very powerful; the incisors, six in number in each jaw, small; the molars, sometimes adapted merely for cutting, in others surmounted with rounded tubercles, presenting no conical points, arranged as in the insectivora. One of these molar teeth is usually much longer and more cutting than the others, and has therefore been called the carnivorous molar tooth; behind these (on each side) are one or two molars, almost flat, and between the carnivorous molar and the canine a variable number of false molars.   The food of the animal, whether exclusively composed of flesh or mixed with other matters, may be judged of by the varying proportions of these cutting or tuberculated molars.

'Animals of this order have generally the toes armed with claws adapted to hold and to tear their prey; usually also they have no collar-bones.'

The following are the characteristics of the genus 'Felis,' and of the species 'Leo':—

'Their jaws are short, and are acted on by muscles of extraordinary strength; their retractile nails, concealed between the toes in a state of repose by means of elastic ligaments, are never blunted.   Their toes are five in number on the anterior limbs, and four on those behind.   Their hearing is exceedingly fine, and the best developed of all their senses. They see well by day and night, but they are not far-sighted; in some the pupil is elongated vertically, in others it is round. They make great use of the organ of smell; they consult it before eating, and often when anything disturbs them. Their tongue is covered with horny and very rough points. Their coat is in general soft and fine, and the surface of the body very sensible to the touch; their whiskers especially seem to be instruments of great sensibility. Though of prodigious vigour, they generally do not attack animals openly, but employ cunning and artifice. They never push their prey to flight, but, watching by the margins of rivers and pools in covert, they spring at once on their victim.

'At the head of this genus stands the lion, measuring frequently twelve feet in length, or over six feet to the setting on of the tail; about three feet in height, and characterised by the square head, the tuft of hair terminating the tail, and in the male by the mane which flows from the head and neck[21].'

The process by which the Zoologist constitutes the Primary Divisions of animal life, and then descends from these to the Species, is distinguished by the same peculiarities as those which we remarked in reviewing the natural classifications of the Botanist. In one step or other of the classification almost every known characteristic of a species will be found. As we descend the series, the characters gain in definiteness and, as a rule, lose in importance. Moreover, even in the higher divisions of the series, numerous characters are used, and those not always of great apparent importance. Thus, that 'the body is generally covered with hairs' is one of the characters of Mammalia.

The student will now be in a position to understand

---

[21] Milne Edwards, § 414.

the rules which may be laid down for the right conduct of a Natural Classification.

I. Not only the lower, but the higher groups of the series should be so constituted as to differ from one another by a multitude of characters. It is only when, as is the case in the primary divisions of Botany and Zoology, we arrive at the same divisions from a variety of different considerations, that we can feel assured that our groups really correspond with distinctions in Nature. It is this *coincidence*, in the higher groups of the series, of divisions formed on different principles, that distinguishes a Natural from an Artificial Classification.

II. The more *important* characters should be selected for the purpose of determining the higher groups. This is called the *principle of the subordination of characters.* But how are we to determine the relative importance of characters? 'We must consider as the most important attributes,' says Mr. Mill[22], 'those which contribute most, either by themselves or by their effects, to render the things like one another, and unlike other things ; which give to the class composed of them the most marked individuality; which fill, as it were, the largest space in their existence, and would most impress the attention of a spectator who knew all their properties but was not specially interested in any.' This account is perfectly true, but it seems to be hardly sufficiently definite. The following criteria may be proposed for the purpose of discriminating between the more and the less important

[22] Mill's *Logic*, Bk. IV. ch. vii. § 2.

properties of natural objects. (1) A character which is found to furnish an invariable index to the possession of certain other characters is of more importance than a character which furnishes no such index. Thus, the internal structure of an animal is of more importance than its size, and the mode of fructification of a plant than the colour of its flowers. (2) Amongst such characters, a character is regarded as of more or less importance, according as it accompanies a greater or smaller number of other differences. Thus, in the classification of animals, the characters from which the classes unguiculata and ungulata are so called are of more importance than the form of the teeth, which is used in distinguishing the orders. For the same reason, the mode of growth of flowering plants (which leads to the distinction of endogenous and exogenous plants) is of far more importance, as a character, than the number of stamens or pistils. Hence, in constituting the higher divisions of a series we must look for those characters which are accompanied by the largest number of differences.

III. The classification should be gradual, proceeding by a series of divisions and subdivisions. When the group to be classified, consists of an enormous number of species, as in the case of animals and plants, the necessity of observing this rule is obvious. To descend at once from the Primary Divisions to, say, Genera and Species, would render the Classification comparatively worthless. The object of a classification being to bring together those groups which resemble each other and to separate those groups which differ from each other, we must take account of degrees of resemblance and difference, so that, as a rule, the number of gradations will increase with the number of groups to be classified. Both in Botany and Zoology, the grand divisions which seem now to be universally recognised are Primary Divisions, or Sub-Kingdoms (embranchements); Classes, Orders, Genera, and Species. Between these various other divisions are interpolated, according to the seeming requirements of each particular system, and often according to the views of each individual author. Moreover, below Species are often reckoned Varieties, and even Varieties are sometimes subdivided, this being especially the case when animals have become domesticated or plants cultivated. Taking as an instance the Anthyllis Vulneraria (Common Lady's Finger), the divisions and subdivisions of a natural classification may be illustrated thus[23]:—

I. PRIMARY DIVISION	. . .	Cotyledones.
II. CLASS	.	Dicotyledones.
Subclass	.	Calyciflorae.
III. ORDER	.	Leguminosae.
Suborder	.	Papilionaceae.
Tribe	.	Loteae.
Subtribe	.	Genisteae.
IV. GENUS	.	Anthyllis.
Subgenus or Section	. .	.. Vulneraria.
V. SPECIES	.	Vulneraria[24].
Variety	.	Dillenii.
Race	.	Floribus coccineis.
Variation	.	Foliis hirsutissimis.

[23] Balfour's *Manual of Botany*, § 725.

[24] It is not uncommon in the classificatory sciences, as in this

In very extensive groups, other divisions may be interpolated; thus a subgenus or section is often divided into a subsection. On the other hand, many of these divisions often disappear; if a genus consist of only a small number of species, and there be no very striking points of difference amongst them, we may descend at once, without any intermediate divisions, from the Genus to its various Species. Sometimes, even, an order may contain only a single genus, or a genus a single species, in which case the two may be regarded as coextensive. In the case of Man, we saw that we descend at once from the Order to the varieties, the Order Bimana being coextensive with the genus and species Homo, so that here three even of the grand divisions are coincident.

IV. The groups should be so arranged, that those which have the closest affinities may be brought nearest to each other, while the distance of one group from another may be taken as a measure of their dissimilarity. The observation of this rule would result in what Mr. Mill calls 'the arrangement of the natural groups into a natural series.' For the purposes of subsequent induction, it is plain that it is of the utmost importance not widely to dissever groups which present many phenomena in common, or which we even suspect may do so. The object aimed at by this rule is, to a great extent, attained by the observation of the Subordination

instance, to assign the same name to a higher and lower division, the lower division exhibiting in a marked manner the characters possessed in common by the various members of the higher division.

of Characters (Rule 2), according to which, the higher the place of the division in the series, the more important is the collection of characters which serves to constitute it. If Rule 2 were duly observed, it would be impossible for any two widely dissimilar groups to be brought into juxtaposition in the lower divisions of the series. Thus, the ox and the frog, the primrose and the mushroom, would in any natural system be at considerable distances from each other. But it is not sufficient to observe the rule of the Subordination of Characters. The arrangement of the cognate groups in each division should be such that at the head of the series may come those groups which are most like the groups of the preceding division, while at the bottom of the series may come those groups which are most like the groups of the subsequent division. Thus, suppose that we have Orders A, B, C, of which B resembles A more than C does, and that A is subdivided into the genera a' a'' a''' b' b'' c; B into the genera m' m'' n o p' p''; C into the genera x' x'' y', y'', y''', z (of which the genera represented by the earlier letters of the alphabet are more akin to each other than those represented by the later, and conversely): in our arrangement we ought to place c in juxtaposition with m' m'', and p' p'' in juxtaposition with x' x'', the remaining groups being arranged, as above, on the same principle. If such an arrangement could be effected, it is plain that those groups which presented in the greatest intensity the principal phenomena of the class of objects under investigation would come first in the series, and that those which presented them in the least intensity would come last. In Zoology, for instance, those groups would come first which presented in the greatest intensity the principal phenomena of animal life, and in Botany those which presented in the greatest intensity the principal phenomena of vegetable life. It is, of course, seldom, in the arrangement of natural objects, that we are able to draw up an exact table of precedence amongst the groups of any division, but we are often able to say that this or that group or collection of groups (a or a' a'' a''') should rank first in the series, or that it should rank before some other group or collection of groups. Thus, no zoologist would hesitate to assign to man (the Order Bimana) the highest place in any classification of Mammalia, while he would place next the Order Quadrumana, and in this Order he would select apes, and, amongst apes, the anthropoid apes, to be brought into closest juxtaposition with man.

This rule is obviously of most difficult application. It points out an ideal to be aimed at, but one which is never likely to be perfectly realised. So many are the properties to be taken into consideration in every natural object, that it is often impossible to say that this object is, on the whole, more like another than that. The groups of the higher divisions may often, those of the lower may sometimes, be tabulated in some order of precedence ; but there remains a large number of cases to which the rule is inapplicable, or to which, at least, it has not yet been successfully applied. This is especially the case in Botany, where, though, in respect of complexity of structure and perfection of organism, Vascular plants may be ranked above Cellular, and Dicotyledons above Monocotyledons, there are many cases among the subdivisions, especially of Monocotyledons and Dicotyledons, where no order of precedence can as yet be satisfactorily established. But, even if the rule were of universal application, and if we were perfectly acquainted with all the properties of bodies and their relative value, it would not follow that we could establish what Dr. Whewell, in his opposition to this doctrine of Classification by Series, calls 'a mere linear progression in nature.' There might still be many Orders, Genera, or Species, which, to use a familiar expression, we should be obliged to *bracket.* 'It would surely be possible,' says Mr. Mill[25], 'to arrange all *places* (for example) in the order of their distance from the North Pole, though there would be not merely a plurality, but a whole circle of places at every single gradation in the scale.'

Remark 1. A natural classification is supposed to be complete, when it has descended as low as species,— a species being regarded as a group consisting of individuals, all of which have descended from a common stock. Or, if the process be reversed, and the classification be an ascending instead of a descending one, species are regarded as the starting-point of the naturalist, and it is supposed that the problem before him is to group them

[25] Bk. IV. ch. viii. § 1. Note.

under higher divisions. But a species may, as we have seen, be divided into varieties, sub-varieties, &c. Now, in what consists the difference between the relation of a variety to a species and the relation of a species to a genus? To this question a very large section of naturalists now maintain that no satisfactory answer can be given. If it be said that varieties of the same species may be developed in the course of time, but that species themselves must be regarded as distinct, it may be asked on what grounds this supposition rests. Different varieties of the same species are certainly more like each other than different species of the same genus, just as species of the same genus have more resemblance than genera of the same order, or members of any lower division than members of any higher division; but, given a larger amount of time, is there more difficulty in supposing a common stock for the different species of a genus than for the different varieties of a species? This is the question originated with so much ability by Mr. Darwin in his work on the *Origin of Species*. His own solution of the question is well known. 'It will be seen,' he says[26], 'that I look at the term species, as one arbitrarily given for the sake of convenience to a set of individuals closely resembling each other, and that it does not essentially differ from the term variety, which is given to less distinct and more fluctuating forms. The term variety, again, in comparison with mere individual differences, is also applied arbitrarily, and for mere

[26] Darwin's *Origin of Species*, ch. ii.

convenience' sake.' It does not fall within my province to discuss the question of the 'Origin of Species,' but it is desirable that the student should be aware that the practice of naturalists in stopping at species, as if they were the 'infimæ species' of the old logicians below which divisions need not proceed, is far from being universally accepted.

Remark 2. As our knowledge of the external world becomes enlarged, the number of natural groups, recognised by the classificatory sciences, is being continually increased. Botanists and zoologists (especially the former) are constantly discovering or recognising new varieties, frequently new species, and occasionally, even, new genera and orders. 'The known species of plants,' says Dr. Whewell[27], 'were 10,000 at the time of Linnæus, and are now [A.D. 1858] probably 60,000.' The increase in the number of recognised varieties, sub-varieties, &c., is even still more rapid. One common effect of these constant discoveries and recognitions is to bridge over what previously appeared to be gaps in nature, thus illustrating the fact that there are but few breaks in natural phenomena, that there pervades nature a Law of Continuity, according to which a group seldom occurs to which some other group may not be found very closely allied. So complete, sometimes, is this continuity, that it becomes very difficult to distinguish the groups by any fixed characters. Two species (say) are discriminated, and then a third group is found which partakes of the character of each of the

[27] *History of Scientific Ideas*, Bk. VIII. ch. ii. § 6.

others. This is constituted a new species, and then a fourth group is found intermediate between this and the first, and so on. 'It has been shown,' says Dr. Carpenter, as quoted by Sir W. Grove[28], 'that a very wide range of variation exists among Orbitolites, not merely as regards external form, but also as to plan of development; and not merely as to the shape and aspect of the entire organism, but also with respect to the size and configuration of its component parts. It would have been easy, by selecting only the most divergent types from amongst the whole series·of specimens which I have examined, to prefer an apparently substantial claim on behalf of these to be accounted as so many distinct species. But after having classified the specimens which could be arranged around these types, a large proportion would yet have remained, either presenting characters intermediate between those of two or more of them, or actually combining those characters in different parts of their fabric; thus showing that no lines of demarcation can be drawn across any part of the series that shall definitely separate it into any number of groups, each characterised by features entirely peculiar to itself.' We certainly find in nature a persistency of type, which is the result of the laws of hereditary transmission; if there were no such persistency, the attempt to group natural objects would be fruitless and absurd. But, at the same time, when we have succeeded in establishing groups,

[28] Essay on Continuity, printed at the end of the Fifth Edition of *The Correlation of Physical Forces*, pp. 326, 327.

we constantly find that there are individual members diverging more or less from the ordinary type, and forming intermediate links between proximate classes. To adopt and alter a metaphor employed by Dr. Whewell, natural classes may be regarded as the forests of neighbouring hills, the hills being seldom separated by well-defined valleys, and the valleys being frequently interspersed with straggling trees or clumps.

Remark 3. It sometimes happens that one of the characters by which classes or groups are distinguished, one from another, is to be found, not invariably, but only usually or occasionally in the members of the group. Thus, in the description of the Order Rosaceæ, we find that 'the seeds are erect or inverted, *usually* exalbuminous. . . . . Flowers sometimes unisexual.' Such indefinite descriptions would be entirely out of place in an artificial classification, but in a natural classification, where the entire assemblage of the characters must be taken into consideration, a character, though not found invariably, or even though found but seldom, may still be valuable in distinguishing a group.

Remark 4. The most important characters are not always those by which a group is most easily recognised. For the purpose of recognition, some external and prominent character or set of characters is generally best adapted. Thus, if we wished to determine whether a plant were monocotyledonous or dicotyledonous, our easiest course would be to examine the stem; if the stem were endogenous, we should know that the plant was a monocotyledon, if exogenous, that the plant was a dicotyledon. A single character is often sufficient to determine the place of a plant or animal in a series, because we already know that the possession of this character is a sign of the possession of the various other characters which are enumerated in the description of the natural class. The method of determining, by means of one or a few characters, the place of a natural object in a classification, is often called <u>Diagnosis</u> or <u>Characteristick</u>. 'The Characteristick,' says Dr. Whewell[29], 'is an Artificial Key to a Natural System. As being Artificial, it takes as few characters as possible ; as being Natural, its characters are not selected by any general or prescribed rule, but follow the natural affinities. 'The genera Lamium and Galeopsis (Dead Nettle and Hemp Nettle) are each formed into a separate group in virtue of their general resemblances and differences, and not because the former has one tooth on each side of the lower lip, and the latter a notch in its upper lip, though they are distinguished by these marks.'

Note.—Dr. Whewell maintains that natural classes are determined, not by *definition*, that is, by an enumeration of characters, but by *type*, that is, by resemblance, more or less complete, to some one member of the class, round which the others are grouped. Thus, according to this theory, the Class Mammalia would be determined, not

[29] *History of Scientific Ideas,* Bk. VIII. ch. ii. § 7.

by an enumeration of characters, but by resemblance, more or less complete, to some typical specimen, say Dog; the genus Dog would be determined not by an enumeration of the characters which are common to the dog, wolf, and fox (the species comprised in the genus), but by approximation to the type-species dog: similarly, the Order Rosaceæ would be determined not by an enumeration of characters, common to a large number of genera, but by the resemblance, more or less complete, of these genera to the type-genus Rosa. Dr. Whewell's view will be understood from the following extract :—

‘In a Natural Group . . . . . . the class is steadily fixed, though not precisely limited; it is given, though not circumscribed; it is determined, not by a boundary line without, but by a central point within; not by what it strictly excludes, but by what it eminently includes; by an example, not by a precept; in short, instead of Definition we have a *Type* for our director.

‘A Type is an example of any class, for instance, a species of a genus, which is considered as eminently possessing the characters of the class. All the species which have a greater affinity with this Type-species than with any others, form the genus, and are ranged about it, deviating from it in various directions and different degrees. Thus a genus may consist of several species, which approach very near the type, and of which the claim to a place with it is obvious; while there may be other species which straggle further from this central knot, and which yet are clearly more connected with it than with any other. And even if there should be some species of which the place is dubious, and which appear to be equally bound by two generic types, it is easily seen that this would not destroy the reality of the generic groups, any more than the scattered trees of the intervening plain prevent our speaking intelligibly of the distinct forests of two separate hills.

'The Type-species of every genus, the Type-genus of every family, is, then, one which possesses all the characters and properties of the genus in a marked and prominent manner. The Type of the Rose family has alternate stipulate leaves, wants the albumen, has the ovules not erect, has the stigmata simple, and besides these features, which distinguish it from the exceptions or varieties of its class, it has the features which make it prominent in its class. It is one of those which possess clearly several leading attributes; and thus, though we cannot say of any one genus that it *must* be the Type of the family, or of any one species that it *must* be the Type of the genus, we are still not wholly to seek : the Type must be connected by many affinities with most of the others of its group; it must be near the centre of the crowd, and not one of the stragglers[30].'

---

[30] *History of Scientific Ideas*, Bk. VIII. ch. ii. § 3. art. 10. Mr. Mill (*Logic*, Bk. IV. ch. vii. §§ 3, 4) examines Dr. Whewell's views at considerable length. He appears to me, in this examination, to insist too emphatically on what he calls ' distinctions of kind,' and to assert, without sufficient warrant, that 'the species of Plants are not only real kinds, but are probably, all of them, real lowest kinds,

There is much force in what Dr. Whewell here says, but his main position appears to me to be incorrect. May not the various steps in the process of Classification be described as follows? We, first, observe a general resemblance amongst a variety of groups. Prompted by the observation of this resemblance, we determine to constitute the groups into a distinct class. But it is not sufficient simply to enumerate the groups which the class contains; it is incumbent upon us to state the principle on which the classification is made. This statement consists in an enumeration of those characters which are common to all the members of the newly-constituted class, and which, at the same time, distinguish them from the members of other classes, with the addition, in some cases, of certain characters which belong to most, or even to a few only, of the members of the class. Thus, the class is *determined* (or '*given*,' to use Dr. Whewell's expression) by an enumeration of characters. But, when the class is once familiar to us, the repetition of the class-name suggests, not the characters, but some typical specimen of the class, some one group which stands out prominently as possessing the characters by which the class was determined; and it is by reference to this central specimen, as it were, that we fix the position of the other groups and adjudicate on the claims of

Infimæ Species, which if we were to subdivide into sub-classes, the subdivision would necessarily be founded on *definite* distinctions, not pointing (apart from what may be known of their causes or effects) to any difference beyond themselves.'

any newly-discovered group to take its place by the side of the others. Thus, the type-species, type-genus, or typical order, may be of the greatest service as a convenient embodiment of the characters, but the characters must be enumerated, and the class determined, before we can select our typical example.

(2) OF NOMENCLATURE.

Nomenclature is intimately connected with Classification. The groups, whether natural or artificial, into which objects are distributed, could neither be recollected by ourselves nor communicated to others, unless they were fixed by the imposition of names. A Nomenclature is a collection of such names, applied to the members of the various divisions and subdivisions which constitute a classification. The number of natural groups, however, is so enormously large, that it would be next to impossible to devise, and, if possible to devise, it would be impossible to remember, distinct names for each group. Thus, the known species of plants, for instance, amount to upwards of 60,000, and, if we took into account varieties, sub-varieties, &c., the number of groups would be represented by many multiples of this sum. Some artifice, therefore, is necessary by which a comparatively small number of names may be made to distinguish a large number of groups. Botany and Chemistry furnish admirable examples of the employment of such an artifice, and some knowledge of the principles which guide the imposition of names in those two sciences (a knowledge which may be easily acquired) would probably be of more service to the student than anything which he might learn from a body of rules for Nomenclature in general.

In Botany, the higher groups (including genera) have distinct names. Thus, we have Dicotyledones, Rosaceæ, Rosa, &c. But, when we arrive at the species, these are known by the generic name with the addition of some distinctive attribute. Thus, the genus Geranium is represented in the British Isles by thirteen species, called respectively Geranium phæum, G. nodōsum, G. sylvaticum, G. pratense, G. sanguineum, G. pyrenaicum, G. pusillum, G. dissectum, G. columbinum, G. rotundifolium, G. molle, G. lucidum, G. robertianum. The specific names are selected from various considerations; sometimes in honour of an individual (as Equisetum Mackaii, Rosa Wilsoni), sometimes from the country or the district in which the plant abounds, sometimes from the soil which is most favourable to it, sometimes from some peculiarity in the plant itself. So arbitrary and fanciful sometimes are these names, that Linnæus (as we are told by Dr. Whewell[31]) 'gave the name of *Bauhinia* to a plant with leaves in pairs, because the Bauhins were a pair of brothers, that of Banisteria to a climbing plant, in honour of Banister, who travelled among mountains.' It is plain that a name which describes some peculiarity in the plant itself is of most service to the learner; but any name, easily remembered, serves

[31] *History of Scientific Ideas*, Bk. VIII. ch. ii. § 6.

the main purpose of a nomenclature, which is to distinguish one group from another. Varieties, sub-varieties, &c., are distinguished from each other on the same principle as species. Thus, as we have seen, of the species Anthyllis Vulneraria there is a variety Dillenii, and of the variety Anthyllis Vulneraria Dillenii there is a 'race' Floribus coccineis, and of the race there is a 'variation' Foliis hirsutissimis. The nomenclature of Zoology is now generally constructed on the same principle as that of Botany. In some systems of Mineralogy, three names are employed, namely, those of the Order, Genus, and Species, as, for instance, Rhombohedral Calc Haloide.

The nomenclature of Chemistry, or, at least, of Inorganic Chemistry, which, in some respects, furnishes an interesting example of a scientific nomenclature, is constructed on the principle of making the prefixes and affixes of the words employed significant of the nature of the substances for which they stand. Thus, we have the affixes *ide, ic, ous, ate, ite,* &c., and the prefixes *mono, di, tri, sesqui,* &c., each having a special significance, though, unfortunately, not always an unambiguous one.

It would transcend the limits of this work to give an account, sufficiently clear and precise, of the Nomenclature of Inorganic Chemistry (which, moreover, is at present in a transitional state), but the student, who is anxious to gain some idea of the principles on which it is constructed, can refer to Watts' *Dictionary of Chemistry,* vol. iv. art. Nomenclature.

(3) Of Terminology.

A Nomenclature of a Science is, as we have seen, a collection of names of groups. A Terminology is a collection of the names (or terms) which distinguish either the properties or the parts of the individual objects which the science recognises. Thus, when we speak of the genus 'Rosa,' we are employing the nomenclature of Botany; but, when we say that the individuals of the genus 'Rosa' have 'their corolla imbricated before flowering, their styles with lateral insertion, their carpels numerous,' &c., we are employing not the nomenclature, but the terminology, of the science. In botany we have an almost perfect example of a complete and judiciously constructed terminology.

'The formation of an exact and extensive descriptive language for botany,' says Dr. Whewell [32], 'has been executed with a degree of skill and felicity, which, before it was attained, could hardly have been dreamt of as attainable. Every part of a plant has been named; and the form of every part, even the most minute, has had a large assemblage of descriptive terms appropriated to it, by means of which the botanist can convey and receive knowledge of form and structure, as exactly as if each minute part were presented to him vastly magnified. This acquisition was part of the Linnæan Reform. "Tournefort," says Decandolle, "appears to have been the first who really perceived the utility of fixing the sense of terms

---

[32] *History of Scientific Ideas*, Bk. VIII. ch. ii. § 2.

in such a way as always to employ the same word in the same sense, and always to express the same idea by the same word; but it was Linnæus who really created and fixed this botanical language, and this is his fairest claim to glory, for by this fixation of language he has shed clearness and precision over all parts of the science."

'It is not necessary here to give any detailed account of the terms of botany. The fundamental ones have been gradually introduced, as the parts of plants were more carefully and minutely examined. Thus the flower was successively distinguished into the *calyx*, the *corolla*, the *stamens*, and the *pistils*: the sections of the corolla were termed *petals* by Columna; those of the calyx were called *sepals* by Necker. Sometimes terms of greater generality were devised; as *perianth* to include the calyx and corolla, whether one or both of these were present; *pericarp* for the part inclosing the grain, of whatever kind it be, fruit, nut, pod, &c. And it may easily be imagined that descriptive terms may, by definition and combination, become very numerous and distinct. Thus leaves may be called *pinnatifid, pinnatipartite, pinnatisect, pinnatilobate, palmatifid, palmatipartite*, &c., and each of these words designates different combinations of the modes and extent of the divisions of the leaf with the divisions of its outline. In some cases arbitrary numerical relations are introduced into the definition: thus a leaf is called *bilobate* when it is divided into two parts by a notch; but, if the notch go to the middle of its length, it is *bifid*; if it go near the base of the leaf, it is *bipartite*; if to the base, it is *bisect*.

Thus, too, a pod of a cruciferous plant is a *siliqua* if it be four times as long as it is broad, but if it be shorter than this it is a *silicula*. Such terms being established, the form of the very complex leaf or frond of a fern is exactly conveyed by the following phrase: "fronds rigid pinnate, pinnæ recurved subunilateral pinnatifid, the segments linear undivided or bifid spinuloso-serrate."'

A Terminology, I have said, comprises the terms appropriated to express, not only the parts of objects, but also their properties. Thus, in the foregoing example, the words 'sepals,' 'petals,' &c., express parts of the plant, the words 'pinnatifid,' 'bilobate,'.&c., which are applied to the shape of the leaves, express characters or properties. A complete terminology must be so constructed as to express every shade of difference in all those properties which are recognised in a scientific treatment of the object. Thus, if colour be regarded as of importance in the description of a plant, mineral, &c., it is essential that there shall be some appropriate term by which to describe every shade of colour. But there are few terms which, from their mere signification, can call up any precise idea in the mind. Hence it is necessary to fix by convention the precise meaning of every technical term employed in science. Again, to appropriate the words of Dr. Whewell, 'The meaning of technical terms can be fixed in the first instance only by convention, and can be made intelligible only by presenting to the senses that which the terms are to signify. The knowledge of a colour by its name can only be taught through the eye. No description can convey to a hearer what we mean by *apple-green* or *French grey*. It might, perhaps, be supposed that, in the first example, the term *apple*, referring to so familiar an object, sufficiently suggests the colour intended. But it may easily be seen that this is not true; for apples are of many different hues of green, and it is only by a conventional selection that we can appropriate the term to one special shade. When this appropriation is once made, the term refers to the sensation, and not to the parts of this term; for these enter into the compound merely as a help to the memory, whether the suggestion be a natural connexion as in "apple-green," or a casual one as in "French grey." In order to derive due advantage from technical terms of this kind, they must be associated *immediately* with the perception to which they belong; and not connected with it through the vague usages of common language. The memory must retain the sensation; and the technical word must be understood as directly as the most familiar word, and more distinctly. When we find such terms as *tin-white* or *pinchbeck-brown*, the metallic colour so denoted ought to start up in our memory without delay or search[33].' When we have fixed, by convention, the meaning of a term, it must invariably be employed in this sense, and in no other. The least vagueness or inconsistency in the use of terms may interpose a fatal obstacle in the way, not only of the learners, but

[33] *History of Scientific Ideas*, Bk. VIII. ch. ii. § 2.

even of the promoters of a science. The progress of the Mechanical Sciences and of what are commonly called Physics was long retarded by the vague and unintelligent use of such words as 'heavy,' 'light,' 'hot,' 'cold,' 'moist,' 'dry,' &c. Even still such words as 'force,' 'fluid,' 'attraction,' 'ether,' &c., are often employed without sufficient precision.

A Terminology, as remarked by Dr. Whewell[34], is indispensably requisite in giving fixity to a Nomenclature. Thus, in Botany, 'the recognition of the kinds of plants must depend upon the exact comparison of their resemblances and differences ; and, to become a part of permanent science, this comparison must be recorded in words.'

Dr. Whewell devotes the last Book of his *Novum Organon Renovatum* to a series of aphorisms on the 'Language of Science,' including both Nomenclature and Terminology. These aphorisms afford one of the best examples of Dr. Whewell's style and mode of treatment, and will well repay the attention of the student who is desirous of acquainting himself further with the methods of the Classificatory Sciences. Mr. Mill has some chapters (*Logic*, Bk. IV. ch. iii–vi) on 'Naming' and the 'Requisites of a Philosophical Language,' and, in addition to the passage already referred to, Dr. Whewell treats these subjects in his *History of Scientific Ideas*, Bk. I. ch. ii ; Bk. VIII. ch. ii. §§ 2 and 6 ; Bk. VIII.

[34] *Novum Organon Renovatum*, Bk. IV. Aphorism ii.

ch. iii. art. 5. In Mr. Bain's *Inductive Logic*, there is
a special chapter (Bk. IV. ch. iii) on Classification, and
another (Bk. V. ch. vi) on the Sciences of Classification.

§ 3. *On Hypothesis.*

When the mind has before it a number of observed
facts, it is almost irresistibly driven to frame for itself
some theory as to the mode of their co-existence or
succession. It is from this irresistible impulse to refer to
some law the various phenomena around us that all
science as well as all scientific error has sprung. In some
cases, as we have seen in the first chapter[35], a single
observation or experiment may at once establish a true
theory or valid induction, independently of any previous
suppositions on our part. But, in all the more intricate
branches of enquiry, true theories have usually been
preceded by a number of false ones, and it has not
unfrequently occurred that the false theories have been
mainly instrumental in conducting to the true. Thus,
the elliptical theory of planetary motion was preceded by
the circular theory, with its various modifications, and
the undulatory theory of light by the emission theory.
Rather than rest satisfied with a number of disconnected
facts, men have often imagined the most absurd relations
between phenomena, such as that a comet was the har-
binger of war, or that the future could be foretold by
birds. These theories, assumptions, or suppositions,
when employed provisionally in scientific enquiry and

[35] See pp. 11, 12.

H

falling short of ascertained truths, are called *hypotheses*, and have already been alluded to in the first chapter. The word 'hypothesis,' as commonly employed, is exclusive of propositions which rest upon absolute proof, whether inductive or deductive, and is generally used in contradistinction to them. Thus, we speak of a science being only in a hypothetical stage, or of a hypothesis being converted into an induction or being brought deductively under some general law already ascertained to be true. On the other hand, we should hardly dignify with the name of 'hypothesis' a supposition which, at least in the eyes of its framer, did not possess some amount of plausibility. A hypothesis [36] may be described as a supposition made without evidence or without sufficient evidence, in order that we may deduce from it conclusions agreeing with actual facts. (If these conclusions are correctly deduced, and really agree with the facts, a presumption arises that the hypothesis is true.) Moreover, if the hypothesis relates to the cause, or mode of production of a phenomenon, it will serve, if admitted, to explain such facts as are found capable of being deduced from it. And this explanation is the purpose of many, if not most, hypotheses. Explanation, in the scientific sense, means the reduction of a series of facts which occur uniformly but are not connected by any known law of causation into a series which is so connected, or the reduction of complex laws of causation

[36] The following sentences, to the end of the paragraph, are slightly altered from Mr. Mill's *Logic*, Bk. III. ch. xiv. § 4.

into simpler laws. If no such laws of causation are known to exist, we may *suppose* or *imagine* a law that would fulfil the requirement ; and this *supposed* law would be a hypothesis.

A hypothesis may be serviceable in many ways. In the first place, it may afford a solution, more or less probable, of a problem which is incapable of any definite solution, or which, at least, has not yet been definitely solved. Thus, many of the advocates of the Darwinian hypothesis maintain that it is the most probable solution of an insoluble problem. Secondly, what was at first started as a hypothesis may ultimately be established by positive proof, as has been the case with the elliptical theory of planetary motion, and, as many suppose, with the undulatory theory of light. Thirdly, even though a hypothesis may ultimately be discovered to be false, it may be of great service in pointing the way to a truer theory. Thus, as already remarked, the circular theory of planetary motion, and the supplementary theory of epicycles and eccentrics, undoubtedly contributed to the formation of the hypothesis which was eventually proved to be true. Kepler himself tried no less than nineteen different hypotheses, before he hit upon the right one, and his ultimate success was doubtless in no slight degree due to his unsuccessful efforts. There is hardly any branch of science in which it might not be affirmed that, without a number of false guesses, true theories could never have been attained. Lastly, a hypothesis, whether true or false, if it be applicable to all the known facts, serves as a means of binding those facts together, of *colligating* them, to use a technical phrase, and thus, by presenting them under one point of view, plainly marks off the phenomena to be explained. A theory, like the Phlogistic theory in Chemistry, or the theory of epicycles and eccentrics (which, by being sufficiently extended, might have been made to include all the phenomena of planetary motion), may thus have been of the greatest service in the history of science, simply by keeping before the minds of investigators the precise phenomena which demanded an explanation.

The formation of hypotheses is obviously the work of the imaginative faculty, a faculty of hardly less importance in science than in art. To lay down rules for the exercise of this faculty has hitherto been found futile. The object of Inductive Logic is rather to restrain the exuberant, than to excite the sluggish, imagination. The latter office is best fulfilled by recounting the great achievements of science, and thus arousing the ambition and kindling the enthusiasm of her votaries. The former (which is no less necessary) may be promoted by determining the conditions to which a hypothesis must conform, in order that it may rank as a provisional explanation of facts, and before it is entitled to demand the honours of a rigorous inductive examination. These conditions may be reduced to three :—

I. The hypothesis must not be known or suspected to be untrue, that is to say, it must not be inconsistent with facts already ascertained or the inferences to which they lead [37]. It would be absurd, for instance, in the present state of knowledge, to propose design or compact as the cause of the divergences which are found in the various dialects of a language, or to suppose the heavenly bodies to move in perfect circles. So simple a rule as this may appear to be superfluous, but it seems necessary to include it in the conditions to which a hypothesis must conform, as, otherwise, a perverted ingenuity might succeed in framing numberless hypotheses which violated none of the preliminary conditions.

II. The hypothesis must be of such a character as to admit of verification or disproof, or at least of being rendered more or less probable, by subsequent investigations [38]. Unless this restriction were placed on the formation of hypotheses, there would be no limit to the wildness of conjecture in which theorists might indulge.

[37] The explanation of this rule, contained in the latter clause of the sentence, has been suggested by Mr. Jevons' chapter on the Use of Hypothesis, a chapter which may be read with advantage by the student. His second condition of a legitimate hypothesis, which corresponds with my first, is expressed thus: ' That it do not conflict with any laws of nature, or of mind, which we hold as true.' *Principles of Science*, vol. ii. p. 139.

[38] It may occur to the student that I have not provided for the case where a supposition is already supported by a certain amount of probable evidence, but where it is not likely to be rendered more or less probable by further investigation. But such a supposition, though it would be an imperfect induction or deduction, could hardly be called a hypothesis, a term which seems always to imply something provisional, something which, on further enquiry, may be either confirmed or weakened, rendered more or less probable than it now is.

It might, for instance, be maintained that falling bodies are dragged to the earth by the action of invisible spirits, and, wild as such a theory would be, there is nothing positively to disprove it. Granted that, like many other products of imagination, such a theory might possibly be true, it would still fall without the scope of science. The aim of science is proof, present or prospective, and consequently what neither admits of proof, nor, so far as we can foresee, is ever likely to admit of it, or even of approximation to it, is no fitting object of scientific enquiry. As affording a caution against the unrestrained exercise of the imagination in scientific speculation, it may be useful to adduce a few instances of .suppositions or hypotheses, which were probably considered as perfectly satisfactory by. those who proposed them or amongst whom they were prevalent, which would now be regarded by all competent authorities as absurd, and which still do not admit of being distinctly disproved.

It was once very generally held that the position of the planets with reference to the earth at any particular moment determines not only the course of human events at that time, but the subsequent life of each person born under the 'conjuncture.' Such an absurd theory is now probably held by no single person of sound understanding; but, so complicated is the web both of society and of individual life, and so easy would it be to explain 'apparent exceptions' by having recourse to 'counteracting causes,' that, if any ingenious person were to maintain and defend this theory, it would probably be impossible to disprove it. Palmistry affords another instance of the same kind. The interlacing of the lines on the palms of the hands is said to indicate a man's 'fortunes.' Such a notion is too absurd to be discussed; but, if maintained, how could it be disproved? It might always be said that the general theory of palmistry was true, though there might be some error in the particular rules by which the 'fortune' in question was foretold [39].

The early history of Geology is full of hypotheses of this kind. The following examples of theories, which no scientific man would now entertain, but which hardly admit of disproof, are extracted from Lyell's *Principles of Geology* [40] :—

'Andrea Mattioli, an eminent botanist, the illustrator of Dioscorides, embraced the notion of Agricola, a skilful German miner, that a certain "materia pinguis," or "fatty

[39] The superstitions connected with dreams afford a similar instance: 'The ancients were convinced that dreams were usually supernatural. If the dream was verified, this was plainly a prophecy. If the event was the exact opposite of what the dream foreshadowed, the latter was still supernatural, for it was a recognised principle that dreams should sometimes be interpreted by contraries. If the dream bore no relation to subsequent events unless it were transformed into a fantastic allegory, it was still supernatural, for allegory was one of the most ordinary forms of revelation. If no ingenuity of interpretation could find a prophetic meaning in a dream, its supernatural character was even then not necessarily destroyed, for Homer said there was a special portal through which deceptive visions passed into the mind, and the Fathers declared that it was one of the occupations of the dæmons to perplex and bewilder us with unmeaning dreams.'—Lecky's *History of European Morals*, vol. i. p. 385.

[40] Lyell's *Principles of Geology*, ch. iii.

matter," set into fermentation by heat, gave birth to fossil organic shapes. Yet Mattioli had come to the conclusion, from his own observations, that porous bodies, such as bones and shells, might be converted into stone, as being permeable to what he termed the "lapidifying juice." In like manner, Falloppio of Padua conceived that petrified shells were gene-rated by fermentation in the spots where they are found, or that they had in some cases acquired their form from "the tumultuous movements of terrestrial exhalations." Although celebrated as a professor of anatomy, he taught that certain tusks of elephants, dug up in his time in Apulia, were mere earthy concretions; and, consistently with these principles, he even went so far as to consider it probable that the vases of Monte Testaceo at Rome were natural impressions stamped in the soil. In the same spirit, Mercati, who published, in 1574, faithful figures of the fossil shells preserved by Pope Sixtus V. in the Museum of the Vatican, expressed an opinion that they were mere stones, which had assumed their peculiar configuration from the influence of the heavenly bodies : and Olivi of Cremona, who described the fossil remains of a rich museum at Verona, was satisfied with considering them as mere "sports of nature." Some of the fanciful notions of those times were deemed less unreasonable, as being some-what in harmony with the Aristotelian theory of spontaneous generation, then taught in all the schools. For men who had been taught, in early youth, that a large proportion of living animals and plants was formed from the fortuitous concourse of atoms, or had sprung from the corruption of organic matter, might easily persuade themselves, that organic shapes, often imperfectly preserved in the interior of solid rocks, owed their existence to causes equally obscure and mysterious.'

'As to the nature of petrified shells, Quirini conceived that, as earthy particles united in the sea to form the shells of mollusca, the same crystallizing process might be effected on the land; and that, in the latter case, the germs of the

animals might have been disseminated through the substance of the rocks, and afterwards developed by virtue of
humidity. Visionary as was this doctrine, it gained many
proselytes even amongst the more sober reasoners of Italy
and Germany; for it conceded that the position of fossil
bodies could not be accounted for by the diluvial theory.'

It has been maintained by theologians, more ardent
than discreet, that all fossils were the creations of the
Devil, whose object was either to mimic the Almighty
or to tempt mankind to disbelieve the Mosaic account
of the creation. Such theories admit of no refutation;
every argument, grounded on the resemblance of fossil
remains to living organisms, shows only more distinctly,
to those who have once embraced the idea, the success
of the alleged agent as a mimic or as an impostor.

Other instances of hypotheses which violate this rule are
afforded by the Vortices of Descartes and the Crystalline
Spheres of the ancient astronomers, both of which were
imagined for the purpose of accounting for the phenomena of planetary motion. Both of these hypotheses
have been subsequently disproved by the free passage
of comets through the spaces supposed to be occupied,
according to the one theory, by the Vortices, according
to the other, by the solid Crystalline Spheres. But at
the time they were first started, there was no reasonable
ground for supposing that, if untrue, they could be disproved, and, what is more important, there was no
possibility of proving them or even rendering them
more probable; they were simply freaks of imagination,
incapable of proof and, to all appearance, of disproof.

Another theory more absurd even than that of the solid crystalline spheres, but which has not, like that, been positively disproved, is the curious hypothesis by which Lodovico delle Colombe endéavoured to reconcile the Aristotelian doctrine that the moon was a perfect body with the recent discoveries of Galileo, who, by the aid of his telescope, had found that its surface was full of hollows, and was consequently charged by his enemies with taking a fiendish delight in distorting the fairest works of nature; the apparently hollow parts, suggested Lodovico, were filled with a pure transparent crystal, and so both the astronomer and the Stagirite were right.

It will be observed that I regard hypotheses as admissible, even though they are not likely ever to be positively proved or disproved, provided that the accumulation of further evidence is likely to render them more or less probable. Between such theories and the theories just exemplified, which are neither supported nor likely to be supported by any evidence whatever, there is the widest difference, and, while the one have no place in Science, the other, I conceive, have a legitimate claim to further consideration. The ideal of Science, it is true, is proof; but, while it can never recognise mere freaks of fancy, it is often compelled to rest content with probabilities. Instances of hypotheses such as I have in view are the Darwinian hypothesis and the Meteoric theory of the repair of Solar Heat, to be noticed presently.

III. The hypothesis must be applicable to the description or explanation of all the observed phenomena, and, if it assign a cause, must assign a cause fully adequate to have produced them. A hypothesis, which does not satisfy this requirement, may be at once rejected. Thus, when the circular theory of planetary motion was found inapplicable to describe several of the phenomena, it was rightly abandoned, and the theory of epicycles and eccentrics, which, though erroneous, was fully adequate to explain all the known phenomena, was substituted for it. One of the most familiar instances of an *inadequate hypothesis* is the theory started by Voltaire, there is little doubt in irony, that the marine shells found on the tops of mountains are Eastern species, dropped from the hats of pilgrims, as they returned from the Holy Land. Such a theory would obviously be inadequate to account (1) for the numbers of the shells, (2) for the fact that they are found imbedded in the rocks, (3) for their existence far away from the tracks of pilgrims, to say nothing of the fact that many of these shells bear no resemblance to recent Eastern species, while none resemble them exactly. The contrast between an *adequate* and an *inadequate* hypothesis is well illustrated by two of the rival hypotheses by which it is attempted to account for the generation and the maintenance of solar heat. These are respectively the Meteoric Theory and the Theory of Chemical Combustion. In speaking of the former theory, Professor Tyndall says[41] :—

[41] *Heat a Mode of Motion,* 3rd ed. §§ 689–693. Sir William Thomson, however, from various considerations, arrived at the con-

'I have already alluded to another theory, which, however bold it may at first sight appear, deserves our serious attention—the Meteoric Theory of the Sun. Kepler's celebrated statement, that "there are more comets in the heavens than fish in the ocean," implies that a small portion only of the total number of comets belonging to our system are seen from the earth. But besides comets, and planets, and moons, a numerous class of bodies belong to our system which, from their smallness, might be regarded as cosmical atoms. Like the planets and the comets, these smaller asteroids obey the law of gravity, and revolve in elliptic orbits round the sun. It is they which, when they come within the earth's atmosphere, and are fired by friction, appear to us as meteors and falling stars.

'On a bright night, twenty minutes rarely pass at any part of the earth's surface, without the appearance of at least one meteor. Twice a year (on the 12th of August and 14th of November) they appear in enormous numbers. During nine hours in Boston, when they were described as falling as thick as snowflakes, 240,000 meteors were observed. The number falling in a year might, perhaps, be estimated at hundreds or thousands of millions, and even these would constitute but a small portion of the total crowd of asteroids that circulate round the sun. From the phenomena of light and heat, and by direct observation on Encke's comet' (the inference from which 'observation,' however, it may be remarked, is very doubtful), 'we learn that the universe is filled by a resisting

clusion that 'the sun's expenditure [of heat], though *originated*, is not *maintained* by mechanical impact; the low rate of cooling and the consequent constancy of the emission being considered by him as due, in great part, to the high specific heat of the matter of the sun.' See Tyndall's Heat, &c., § 701. Other physicists (see Young on the Sun, pp. 270–7) conjecture that the heat of the sun is partly due to its gradual contraction and the increase of temperature thus generated.'

medium, through the friction of which all the masses of our system are drawn gradually towards the sun. And though the larger planets show, in historic times, no diminution of their periods of revolution, it may be otherwise with the smaller bodies. In the time required for the mean distance of the earth to alter a single yard, a small asteroid may have approached thousands of miles nearer to the sun.

'Following up these reflexions, we should be led to the conclusion that, while an immeasurable stream of ponderable meteoric matter moves unceasingly towards the sun, it must augment in density as it approaches its centre of convergence. And here the conjecture naturally rises, whether that vast nebulous mass, the Zodiacal Light, which embraces the sun, may not be a crowd of meteors. It is at least proved that this luminous phenomenon arises from matter which circulates in obedience to planetary laws; hence, the entire mass of the zodiacal light must be constantly approaching, and incessantly raining its substance down upon the sun.

'It is easy to calculate both the maximum and the minimum velocity, imparted by the sun's attraction to an asteroid circulating round him. The maximum is generated when the body approaches the sun from an infinite distance; the *entire pull* of the sun being then exerted upon it. The minimum is that velocity which would barely enable the body to revolve round the sun close to his surface. The final velocity of the former, just before striking the sun, would be 390 miles a second; that of the latter 276 miles a second. The asteroid, on striking the sun, with the former velocity, would develope more than 9000 times the heat generated by the combustion of an equal asteroid of solid coal; while the shock, in the latter case, would generate heat equal to that of the combustion of upwards of 4000 such asteroids. It matters not, therefore, whether the substances falling into the sun be combustible or not; their being combustible would not add sensibly to the tremendous heat produced by their mechanical collision.

'Here, then, we have an agency competent to restore his lost energy to the sun, and to maintain a temperature at his surface which transcends all terrestrial combustion. In the fall of asteroids we find the means of producing the solar light and heat. It may be contended that this showering down of matter necessitates the growth of the sun: it does so; but the quantity necessary to maintain the observed calorific emission for 4000 years, would defeat the scrutiny of our best instruments. If the earth struck the sun, it would utterly vanish from perception; but the heat developed by its shock would cover the expenditure of a century.'

Of the other theory, Professor Tyndall says [42]:—

'Sir William Thomson adduces the following forcible considerations to show the inadequacy of chemical combination to produce the sun's heat. "Let us consider," he says, "how much chemical action would be required to produce the same effects. . . . Taking the former estimate, 2781 thermal units [43] centigrade (each 1390 foot pounds) or 3,869,000 foot pounds, which is equivalent to 7000 horse-power, as the rate per second of emission of energy from every square foot of the sun's surface, we find that more than 0.42 of a pound of coal per second, 1500 lbs. per hour, would be required to produce heat at the same rate. Now if all the fires of the whole Baltic fleet (this was written in 1854) were heaped up and kept in full combustion over one or two square yards of surface, and if the surface of a globe all round had every square yard so occupied, where could

[42] *Heat a Mode of Motion*, § 700.

[43] The *thermal unit* is the quantity of heat necessary to raise the temperature of a pound of water one degree. If the degree be centigrade, this is equivalent to the heat generated by a pound weight falling from a height of 1390 feet against the earth. The term *foot-pound* expresses the energy requisite to lift one pound to the height of a foot.

a sufficient supply of air come from to sustain the combustion? Yet such is the condition we must suppose the sun to be in, according to the hypothesis now under consideration. ... If the products of combustion were gaseous, they would, in rising, check the necessary supplies of fresh air; if they were solid and liquid (as they might be if the fuel were metallic), they would interfere with the supply of elements from below. In either or in both ways, the fire would be choked, and I think it may be safely affirmed that no such fire could keep alight for more than a few minutes, by any conceivable adaptation of air and fuel. If the sun be a burning mass it must be more analogous to burning gunpowder than to a fire burning in air; and it is quite conceivable that a solid mass, containing within itself all the elements required for combustion, provided the products of combustion are permanently gaseous, could burn off at its surface all round, and actually emit heat as copiously as the sun. Thus, an enormous globe of gun-cotton might, if at first cold, and once set on fire round its surface, get to a permanent rate of burning, in which any internal part would become heated sufficiently to ignite, only when nearly approached by the burning surface. It is highly probable indeed that such a body might for a time be as large as the sun and give out luminous heat as copiously, to be freely radiated into space, without suffering more absorption from its atmosphere of transparent gaseous products than the light of the sun actually does experience from the dense atmosphere through which it passes. Let us therefore consider at what rate such a body, giving out heat so copiously, would burn away; the heat of combustion could probably not be so much as 4000 thermal units per pound of matter burned, the greatest thermal equivalent of chemical action yet ascertained falling considerably short of this. But 2781 thermal units (as found above) are emitted per second from each square foot of the sun; hence there would be a loss of about 0.7

of a pound of matter per square foot per second . . . or a layer half a foot thick in a minute, or 55 miles thick in a year. At the same rate continued, a mass as large as the sun is at present would burn away in 8000 years. If the sun has been burning at that rate in past time he must have been of double diameter, of quadruple heating power, and of eight-fold mass only 8000 years ago. We may therefore quite safely conclude that the sun does not get its heat by chemical action . . . and we must therefore look to the meteoric theory for fuel."'

A hypothesis which fulfils these three conditions is a *legitimate hypothesis*, though it must conform to still more rigorous requirements before it can be accepted as a complete Induction, or even be regarded as possessing any great amount of probability. Thus, the Meteoric Theory, though it is not yet *proved*, and perhaps never may be proved, to be the true explanation of the phenomenon of solar heat, is perfectly tenable as a hypothesis. For, to take the conditions in the reverse order to that in which they have been enumerated above, the impact of a large number of meteors on a body of considerable density, such as the sun probably is, would be competent or adequate to produce the given effect; the theory in question is likely, if not to be proved or disproved, at least to be rendered more or less probable by the progress of astronomical science; lastly, we do not know, nor have we any reason to suppose, that the hypothesis is an untrue explanation of the facts. But, though legitimate as a hypothesis, before we could accept the Meteoric Theory as a Valid or Complete Induction, that is to say, an ascertained truth, we should require to know not only that there is a large number of meteors circulating round the sun, that these meteors have a tendency to fall into the central body, and that, *if* they were falling or had fallen in sufficient quantities, they would be competent or would have been competent to produce the present amount of solar heat, but also that they do, as a matter of fact, fall in sufficient quantities to account for the phenomenon, or, at least, that *nothing else* but the showering down of asteroids and meteors *could* account for it.

It was by availing himself of the latter mode of proof that Newton demonstrated the existence in the sun of a central force attracting the planets towards it. Assuming Kepler's hypothesis (then sufficiently verified by observation to be universally accepted as a true statement of the facts), that equal areas are described by the radii vectores of the planets in equal times, Newton showed that this fact could be due to only one cause, namely, the deflexion of the planets from their rectilinear course by a force acting in the direction of the sun's centre. The existence of the central force was, at first, started by him as a hypothesis. 'He then proved that,' on the supposition of the existence of such a force, 'the planet will describe, as we know by Kepler's first law that it does describe, equal areas in equal times; and, lastly, he proved that if the force acted in any other direction whatever, the planet would not describe equal areas in equal times. It being thus shown that no other hypothesis would accord with the facts, the assumption was proved; the hypothesis became an inductive truth. Not only did Newton ascertain by this hypothetical process the direction of the deflecting force; he proceeded in exactly the same manner to ascertain the law of variation of the quantity of that force. He assumed that the force varied inversely as the square of the distance; showed that from this assumption the remaining two of Kepler's laws might be deduced; and finally, that any other law of variation would give results inconsistent with those laws, and inconsistent, therefore, with the real motions of the planets, of which Kepler's laws were known to be a correct expression [44].'

It will be noticed that the course of demonstration pursued in this instance is the following: (1) we have certain observed facts; (2) these observed facts are 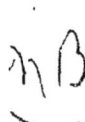 generalised in what are called Kepler's Laws; (3) we have the assumption of the central force; (4) it is shown that the central force will account for Kepler's Laws, and therefore, of course, for the particular facts of observation on which those Laws were founded; (5) it is shown (and this, together with the next step, is what properly constitutes the demonstration) that this assumption is the only one which will account for the Laws or the particular facts expressed by them; (6) it is inferred inductively, by means of the Method of Difference (to be hereafter described), that the assumption of the central force, as it will account for, and is the only supposition

[44] Mill's *Logic,* Bk. III. ch. xiv. § 4.

which will account for, the observed facts, must be accepted as true ; (7) Kepler's Laws (which had hitherto been accepted as correct statements of observed facts, though they had not yet been explained by reference to any cause competent to account for them) are now proved deductively from what we have ascertained to be the Valid Induction of the Central Force.

A Hypothesis can only be converted into a Valid Induction [45] by the application of one or other of the Inductive Methods (to be described in the next Chapter), or, if we insist on strict accuracy of proof, of such of them as furnish absolutely certain conclusions.

Note 1.—According to the view here taken, which agrees with that of Mr. Mill, a hypothesis cannot claim to be regarded as an established truth, till it has conformed to the requirements of one or other of the inductive methods, or has been shown to admit of being deduced from some previously established Induction. Thus, when Newton proves the existence of a central force, deflecting the planets from the recti-

[45] Though a hypothesis is usually contrasted with a Valid or Complete Induction, it must not be forgotten that we have admitted, as legitimate, hypotheses which are never likely to rest on more than probable evidence. These can, of course, receive accessions of proof only by the same means as those by which we establish Imperfect Inductions. It should also be remembered that the truth of a hypothesis may be demonstrated by deductive as well as by inductive methods.

lineal course which they would otherwise describe and making them describe curves round the sun, by showing that no other supposition would account for the fact that their radii vectores describe equal areas in equal times, he is, as Mr. Mill says, employing the Method of Difference. The demonstration 'affords the two instances, A B C, *a b c,* and B C, *b c.* A represents central force; A B C, the planets *plus* a central force; B C, the planets as they would be without a central force. The planets with a central force give *a* (areas proportional to the times); the planets without a central force give *b c* (a set of motions) without *a,* or with something else instead of *a.* This is the Method of Difference in all its strictness. It is true, the two instances which the method requires are obtained in this case, not by experiment, but by a prior deduction. But that is of no consequence. It is immaterial what is the nature of the evidence from which we derive the assurance that A B C will produce *a b c,* and B C only *b c;* it is enough that we have that assurance. In the present case, a process of reasoning furnished Newton with the very instances, which, if the nature of the case had admitted of it, he would have sought by experiment[46].'

Dr. Whewell, who does not acknowledge the utility of Mr. Mill's methods, appears to regard the inductive process as consisting simply in the framing of successive hypotheses, the comparison of these hypotheses with the ascertained facts of nature, and the introduction into them of such modifications as that comparison may

[46] Mill's *Logic,* Bk. III. ch. xiv. § 4.

render necessary [47]. The first requisite in a hypothesis, according to Dr. Whewell, is that it shall explain all the observed facts. But its probability, he urges, will be considerably enhanced, if, in addition to explaining observed facts, it enables us to predict the future. 'Thus the hypotheses which we accept ought to explain phenomena which we have observed. But they ought to do more than this: our hypotheses ought to *foretel* phenomena which have not yet been observed; at least all phenomena of the same kind as those which the hypothesis was invented to explain. For our assent to the hypothesis implies that it is held to be true of all particular instances. That these cases belong to past or to future times, that they have or have not already occurred, makes no difference in the applicability of the rule to them. Because the rule prevails, it includes all cases; and will determine them all, if we can only calculate its real consequences. Hence it will predict the results of new combinations, as well as explain the appearances which have occurred in old ones. And that it does this with certainty and correctness, is one mode in which the hypothesis is to be verified as right and useful [48].'

---

[47] A theory of Induction almost identical with that of Dr. Whewell (though, I venture to suggest, not so clearly stated or so carefully guarded) has been recently propounded by Professor Stanley Jevons in his *Principles of Science.* This theory, together with other points of difference between Professor Jevons and myself, I have noticed in the Preface to the third edition, reprinted in the present one.

[48] *Novum Organon Renovatum,* Bk. II. ch. v. art. 10.

. Curiously enough, the first hypothesis which Dr. Whewell cites, as having fulfilled both these conditions, is also one which eventually proved to be false. 'For example, the Epicyclical Theory of the heavens was confirmed by its *predicting* truly eclipses of the sun and moon, configurations of the planets, and other celestial phenomena; and by its leading to the construction of Tables by which the places of the heavenly bodies were given at every moment of time. The truth and accuracy of these predictions were a proof that the hypothesis was valuable, and, at least to a great extent, true; although, as was afterwards found, it involved a false representation of the structure of the heavens.' A theory may thus not only enable us to explain known facts, but even to predict the future, and still be untrue. Notwithstanding, however, the infelicitous character of the example selected, Dr. Whewell attaches the greatest importance to the fulfilment of this condition by a hypothesis. 'Men cannot help believing that the laws laid down by discoverers must be in a great measure identical with the real laws of nature, when the discoverers thus determine effects beforehand in the same manner in which nature herself determines them when the occasion occurs. Those who can do this must, to a considerable extent, have detected nature's secret;—must have fixed upon the conditions to which she attends, and must have seized the rules by which she applies them. Such a coincidence of untried facts with speculative assertions cannot be the work of chance, but implies some large portion of truth

in the principles on which the reasoning is founded. To
trace order and law in that which has been observed, may
be considered as interpreting what nature has written
down for us, and will commonly prove that we understand her alphabet. But to predict what has not been
observed, is to attempt ourselves to use the legislative
phrases of nature; and when she responds plainly and
precisely to that which we thus utter, we cannot but suppose that we have in a great measure made ourselves
masters of the meaning and structure of her language.
The prediction of results, even of the same kind as those
which have been observed, in new cases, is a proof of real
success in our inductive processes.'

But what appears to Dr. Whewell to establish the truth
of a hypothesis beyond all question is what he calls a
Consilience of Inductions. 'We have here spoken of the
prediction of facts *of the same kind* as those from which
our rule was collected. But the evidence in favour of our
induction is of a much higher and more forcible character
when it enables us to explain and determine cases of
a *kind different* from those which were contemplated in
the formation of our hypothesis. The instances in which
this has occurred, indeed, impress us with a conviction
that the truth of our hypothesis is certain. No accident
could give rise to such an extraordinary coincidence. No
false supposition could, after being adjusted to one class
of phenomena, exactly represent a different class, where
the agreement was unforeseen and uncontemplated. That
rules springing from remote and unconnected quarters

should thus leap to the same point, can only arise from *that* being the point where truth resides.

'Accordingly the cases in which inductions from classes of facts altogether different have thus *jumped together*, belong only to the best established theories which the history of science contains. And, as I shall have occasion to refer to this peculiar feature in their evidence, I will take the liberty of describing it by a particular phrase; and will term it the *Consilience of Inductions*.

'It is exemplified principally in some of the greatest discoveries. Thus it was found by Newton that the doctrine of the Attraction of the Sun varying according to the Inverse Square of the distance, which explained Kepler's *Third Law*, of the proportionality of the cubes of the [mean] distances to the squares of the periodic times of the planets, explained also his *First* and *Second Laws*, of the elliptical motion of each planet; although no connexion of these laws had been visible before. Again, it appeared that the force of Universal Gravitation, which had been inferred from the *Perturbations* of the moon and planets by the sun and by each other, also accounted for the fact, apparently altogether dissimilar and remote, of the *Precession of the equinoxes*. Here was a most striking and surprising coincidence, which gave to the theory a stamp of truth beyond the power of ingenuity to counterfeit[49].'

It is undeniable that a theory which thus appears to afford an explanation of different classes of facts strikes

[49] *Novum Organon Renovatum*, Bk. II. ch. v. art. 11.

the imagination with considerable force, and that its very simplicity furnishes *primâ facie* evidence of its truth. But what is required before a hypothesis can be placed beyond suspicion is formal proof, and that, it appears to me, is furnished by Mr. Mill's 'methods,' and not by Dr. Whewell's requisitions of *explanation, prediction,* and *consilience of inductions.* For the questions at issue between Mr. Mill and Dr. Whewell, see Whewell's *Novum Organon Renovatum* (where his views are stated in their latest and most matured form), Bk. II. ch. v. § 3, and Mill's *Logic,* Bk. III. ch. xiv. § 6.

Note 2.—In attempting to determine the conditions to which a legitimate hypothesis must conform, I have avoided the employment of the expressions *vera causa* and *adæquata causa.* In the first place, a hypothesis may simply attempt to find a general expression for a number of isolated facts without referring them to any cause, as was the case with the various hypotheses respecting the shape of the planetary orbits, and hence to speak as if a hypothesis always assigned a cause is an undue limitation of the meaning of the word. But to the expression *vera causa* there is a more special exception. Its meaning is ambiguous. Is it the actual cause which produces a phenomenon, or a cause which we know to be actually existent, or a cause analogous to an existent cause? The student will find a criticism of this expression (first employed by Newton) in Dr. Whewell's *Philosophy of Discovery,* ch. xviii. § 5, &c. The expression cannot have been used in the first, which is its most obvious, sense, for, as Dr. Whewell says, 'although it is the philosopher's aim to discover such causes, he would be little aided in his search of truth, by being told that it is truth which he is to seek.' But in the second of the two remaining senses, the requirement, as would now be generally acknowledged, is too stringent, and, if it had been invariably observed, would have prevented us from reaping some of the greatest discoveries in science, while in the last it is so vague as to be of no practical service. It has been attempted to affix other meanings to the phrase ; but there can be little doubt that Newton, having in mind the Vortices of Descartes, intended to protest against the introduction of causes of whose existence we have no direct knowledge, and consequently laid down a rule, which the subsequent history of science has shown to be needlessly stringent.

Note 3.—We sometimes find the expression a '*gratuitous* hypothesis.' By this phrase is meant the assumption of an unknown cause, when the phenomenon is capable of being explained by the operation of known causes, or the introduction of an extraneous (though it may be known) cause, when the phenomenon is capable of being accounted for by the causes already known to be in operation. Of the latter case we should have instances, where a man is supposed to have acted at the suggestion of another, though his own motives would supply a sufficient explanation of his conduct, or where a man is supposed to have been poisoned, though he was already known to have been suffering from a fatal disease. Of the former case we should have instances in the crystalline spheres of the ancient astronomers and in the masses of crystal which were supposed by Lodovico delle Colombe to fill up the cavities of the moon (there being no instances known to us of the existence of crystal in these huge masses, and the phenomena being capable of explanation without making the supposition); in the caloric (which was supposed to be a distinct substance) of the early writers on heat; in the 'electrical fluid' of the early electricians; and in the ἀπόρροιαι of Democritus or the 'intentional species' of the Peripatetics, which, being invented for the purpose of explaining the perception of material objects by the mind, were themselves equally in need of explanation. In all these instances, under whichever of the two cases they may fall, the objection to the hypothesis is that it seems 'not to be needed.'

I have said nothing of 'gratuitous hypotheses' in the text, as a hypothesis, though it may appear to be gratuitous, may still be legitimate, and may even ultimately turn out to be true.

CHAPTER III.

On the Inductive Methods.

INDUCTION has been defined to be a legitimate inference from the known to the unknown. But the unknown must not be entirely unknown. It must be known to agree in certain circumstances with the known, and it is in virtue of this agreement that the inference is made. Now, how are we to ascertain what are the common circumstances which justify the inductive inference? X and Y may both agree in exhibiting the circumstances a, b, c, but it will not follow because X exhibits the quality m, that therefore this quality will also necessarily be found in Y. Nor even, if twenty, thirty, a hundred, or a thousand cases could be adduced in which the circumstances a, b, c were found to be accompanied by the circumstance m, would it follow necessarily (it might not even follow probably) that the next case in which we detected the circumstances a, b, c would also exhibit the quality m. We might pass through a field containing thousands of blue hyacinths, but this fact would not justify us in expecting that the next time we saw a hyacinth, it would be a blue one. This form of induction (*Inductio per Enumerationem Simplicem*) may have no value whatever. In most cases, the condemnation passed on it by Bacon[1] is perfectly just: 'Inductio quæ procedit per enumerationem simplicem, res puerilis est, et precario concludit, et periculo exponitur ab instantia contradictoria, et plerumque secundum pauciora quam par est, et ex his tantummodo quæ præsto sunt, pronunciat.' But when we have reason to think that any instances to the contrary, if there were such, would be known to us, the argument may possess considerable value, and when, as in the case of the Laws of Causation and of the Uniformity of Nature, we feel certain, from a wide and uncontradicted experience, that there are no cases to the contrary, no stronger argument (to us individually) can be adduced. It is not often, however, that an Inductio per Enumerationem Simplicem can afford us this certainty[2]. Our trustworthy inductions are, in the majority of cases, the result of our detecting some fact of causation among the observed phenomena. We find, for instance, that, amongst the observed phenomena, a, b, c, d of X, a is the cause of c, and, consequently, if we observe the phenomenon a in Y, we infer that, if there are no counteracting circumstances, Y will possess the quality c as well; or, if we

---

[1] *Novum Organum*, Lib. I. aph. cv.

[2] It must be remembered that a *complete* enumeration of instances, when we know the enumeration to be complete, inasmuch as it leaves no room for an inference from the known to the unknown, does not furnish an inductive but a deductive argument. See *Elements of Deductive Logic*, Part III. ch. i. appended Note 2.

observe the phenomenon c in Y, we infer that it is not unlikely[3] that a may be present as well. The problem of Induction, therefore, resolves itself (except in the cases in which we may legitimately employ Inductio per Enumerationem Simplicem, or the cases in which we have no other resource) into the problem of detecting facts of Causation. Certain rules for this purpose have been laid down by Mr. Mill, called by him the Experimental Methods, but which I shall describe as the Inductive Methods.

These Methods, it will be noticed as we proceed, are all *methods of elimination*, or *devices* by which we are enabled to argue from a comparatively small number of instances with the same certainty as if they were ever so numerous.

Before proceeding to state and explain these Rules or Methods, it may be useful to make some preliminary remarks on the nature of the causal relations which subsist among phenomena.

(1) The same cause, unless there are counteracting circumstances, that is, other causes which prevent it from acting or which modify its action, is invariably followed by the same effect.

(2) As already shown (Chapter I. pp. 13–16), several causes may have co-operated in producing any given effect. In this case, it is not unusual to speak of the 'combination of causes' or the 'sum of the causes.'

---

[3] I say 'not unlikely,' for c might be due to some other cause as well as a, and, therefore, the presence of c does not enable us to infer with certainty the presence of a, as does that of a the presence of c.

(3) The same effect may be due to several distinct causes, or combinations of causes, being due sometimes to one and sometimes to another, and, hence, though we may always argue from a particular cause to its effect, we cannot always argue from an effect to any particular cause. Thus, ignition may be due, not only to the concentration of the rays of solar heat, but also to friction, electricity, &c. This fact has given occasion to the expression '*Plurality of Causes*,' for which a recent writer (Mr. Carveth Read) has proposed to substitute the expression 'Vicariousness of Causes,' in order to distinguish clearly the case of *alternative* from that of *co-operating* or *concurrent* causes, noticed in the last paragraph [4].

[4] It is sometimes doubted whether the same effect is ever really due to different causes, and it may be conceded, I think, that different causes never do produce precisely the same aggregate of effects. Together with certain common effects, they produce certain divergent effects, and it is the presence of these, indeed, that enables us to determine the particular cause which has been at work in the particular instance. There is, however, nothing in this circumstance inconsistent with the occurrence of some one or more effects common to all the causes. Thus, the whole group of effects produced severally by heat, electricity, impact, differs widely, but, at the same time, the motion of a needle may be a common part of the effect in all three instances, and, when we see the needle in motion, we may be unable to say to which of the three causes motion is due. Similar considerations may be applied to the cases of ignition, and death, which are favourite illustrations of the operation of a plurality of causes. Taking A, B, C, D as causes or combinations of causes, and a, b, c, &c. as individual portions of the aggregate effects produced by the causes, we may conceive A as producing $a\ b\ c\ d\ e$, B as pro-

(4) It frequently happens that between the original cause and the ultimate effect there intervene a number of intermediate causes. Thus, suppose we make an experiment by which motion is converted into heat, heat into electricity, and electricity into chemical affinity ; we may, roughly speaking, say that motion has been the cause of the chemical affinity, or chemical affinity the effect of the motion, but, speaking strictly, we ought to enumerate the intervening causes.

(5) Sometimes a number of effects appear to be produced simultaneously by the same cause. Thus, it would appear that there are many cases in which, if one of the agents, motion, heat, light, electricity, magnetism, and chemical affinity, is excited, the rest are developed simul-

ducing *d e f g h,* C as producing *c d e i k l,* D as producing *e m n o.* In this case, *e* may be regarded as an effect due to any one of the causes A, B, C, D, though the 'attendant circumstances,' as they are often called, are widely different in each instance. If, therefore, we were to state the doctrine of Plurality or 'Vicariousness' of Causes exactly, we should say, not that the same effect may be due to different causes, but that, of the total effects due to different causes, a certain portion is often found to be common to all. For purposes of practice, however, the ordinary mode of statement is sufficiently precise.

It seems hardly necessary to remark that it is no valid objection to the doctrine of Plurality of Causes that we are sometimes able to detect between the alternative causes and the identical effect some set of conditions which is the same in all cases. This discovery only removes the plurality of causation one step further back, and the doctrine can only be consistently denied by those who maintain that at no single point in the series of receding causes can we find the same effect produced, or capable of being produced, by distinct causes.

taneously [5]. These simultaneous effects, whether we conceive that they are really or only apparently simultaneous, would be called *joint* or *common* effects of the cause. Similarly the expression 'joint effects' would be employed for the effects produced by the same cause on different bodies, or different portions of the same body. Thus, if a blow bruises my forehead, and at the same time gives me a headache, the bruise and the headache may be called joint effects of the blow. These joint effects may be, as it were, in different degrees of descent from the same cause. Thus, if the headache incapacitates me for work, my incapacity for work and the bruise on my forehead will be joint effects, but in different degrees of descent from the original cause.

Any phenomena which are connected, either as cause and effect, and that either immediately or remotely, or as joint effects, and that either in the same or in different degrees of descent from the same cause, may be spoken of as being *causally connected*, or as *causal relations*, or as being *related* to one another through some *fact of causation*.

I now proceed to the statement of the Inductive Methods.

---

[5] See Grove's *Correlation of Physical Forces*, Concluding Remarks. What Sir W. Grove calls 'Force' would now be denominated 'Energy,' and the doctrine of the 'Correlation of Physical Forces' would be subsumed under that of the 'Conservation of Energy.'

K

METHOD OF AGREEMENT.

CANON [6].

If two or more instances of the phenomenon under investigation have only one other circumstance in common, that circumstance may be regarded, with more or less of probability, as the cause (or effect) of the given phenomenon, or, at least, as connected with it through some fact of causation.

Wherever the phenomenon a is found, we observe that b is found, either invariably or frequently [7], in conjunction with it. This fact leads us to suspect that there is some causal connexion between them. On what grounds, and under what circumstances, are we justified in drawing such an inference? And what is the particular character of the inference which we are justified in drawing? The answer to these questions involves many difficulties, of which I shall now attempt to offer a solution.

When antecedents and consequents are discriminated in this discussion, antecedents will be represented by Roman capitals, A, B, C, &c., and consequents by Greek characters, a, β, γ, &c. When circumstances are not distinguished as antecedents and consequents, I shall employ the small Roman letters, a, b, c, &c.

[6] The statement of the Canons is taken, with some modifications, from Mr. Mill's *Logic*. The authorities for the various examples, when these are not of a familiar character, are cited at the foot of the page.

[7] I add 'or frequently,' as it is not necessary that the conjunction shall be invariable. The student need not, however, at present trouble himself with this distinction, which will be fully explained below. See pp. 137–8, 145–7.

Now, suppose that we have A B followed by $a \beta$, and A C by $a \gamma$; it might, at first sight, appear that A must be the cause of a, or, if we were attempting to ascertain the effect of a given cause (which, however, is a much rarer application of this method), that a must be the effect of A. And there is much plausibility in this supposition, for, provided that all the other circumstances remain the same, whatever can, in any given instance, be excluded, or, to use the technical term, *eliminated* without prejudice to a phenomenon, cannot have any influence on it in the way of causation, nor, making the same proviso, can an effect which disappears be due to a cause which continues to operate. Thus, if we were attempting to find the cause of a given effect a, it might be argued that B cannot be its cause, for it is absent in one of the cases where a is present, and similarly of C; but that a must be due to some cause; and, consequently, it is due to A, the only antecedent remaining. Or, if we were attempting to find the effect of a given cause A, it might be argued that β cannot be its effect, for it is absent in one of the cases where A is present, and similarly of γ; but that, as a has been permanently present, A must be its cause. If it were not for the fact that the same event may be due to a great number of distinct causes (as is exemplified in the familiar cases of motion, death, disease, &c.), this reasoning would be perfectly just. Now it will be observed that, when B was removed, it was replaced by C. It is, therefore, conceivable that a may have been due to B in the first instance, and

K 2

to C in the second, it being, of course, in each case, only a portion of the effect, the remaining portions being respectively β, γ, and A having been throughout inoperative. This consideration, it is plain, vitiates the reasoning, whether we are attempting to discover the effect of a given cause or the cause of a given effect. Thus, suppose that there are two distinct drugs, either of which is potent to remove a given disease, and that, in administering each of them, we mix it with some perfectly inert substance, which is the same in each case; if the principles of the above reasoning were correct, and we were justified in neglecting to take account of what may be called the Plurality of Causes, we should be at liberty to argue (if we were seeking the cause of a given effect) that the restoration of the patients to health was, in each case, due to the inert substance, or (if we were seeking the effect of a given cause) that the inert substance was the cause of their restoration to health.

But, if the Method of Agreement is open to so serious an objection, it may be asked on what grounds is it recognised as an Inductive Method? The answer is that, by the multiplication and variation of instances, the possible error due to the Plurality of Causes may be rendered less and less probable, till, at last, for all practical purposes, it may be regarded as having disappeared. Thus, if to the instances A B, $a\beta$; A C, $a\gamma$; we can add A D, $a\delta$; A E, $a\epsilon$, &c. &c. ; it is plain that we may, at each step, be very considerably diminishing the possibility of an error in our reasoning, and, after a certain number of instances, may be justified in feeling morally certain that we have avoided it. It is not likely that, in a number of instances, each agreeing in some one circumstance (besides the phenomenon which is being investigated) but differing as widely as possible in all other circumstances, the same event should in each case, or in a majority of cases, or in even a great number of cases, be due to different causes. The chance of an inert substance being successively mixed with two potent drugs, and of the effects which are really due to them being erroneously ascribed to it, is, in the present state of medical science, but a very slight one; but the probability is obviously considerably diminished, if instead of two such errors we suppose three, instead of three we suppose four, and so on.

For the sake of simplicity, I have assumed groups of two antecedents and two consequents (A B, $\alpha\beta$; A C, $\alpha\gamma$; &c. &c.), but it is extremely seldom that we find in nature combinations so simple. We have usually a vast mass of antecedents and a vast mass of consequents (or, to state the same proposition in more scientific language, a vast mass of antecedents all, or most of them, contributing to a complex effect), and hence it often becomes a matter of extreme difficulty to discover a collection of instances which, presenting the phenomenon in question, agree in only one other circumstance or even in a small number of other circumstances. The difficulty, therefore, of rigidly satisfying the requirements of the Method must be added to what Mr. Mill calls its *characteristic imperfection*, namely, the uncertainty attaching to its conclusions from the consideration of the Plurality of Causes.

But there is still a third difficulty incident to the Method of Agreement, which however is, in a majority of cases, of a theoretical rather than a practical nature. If we insisted literally on the fulfilment of the condition that the instances presenting the given phenomenon should have only one other circumstance in common, it would be simply impossible to find such instances. All instances will be found to agree in a number of circumstances which are immaterial to the point under investigation. Thus, if we are enquiring into the properties of a group of external objects, they will all agree in the fact that they are subject to the action of gravity, and probably also in the facts that they are surrounded by atmospheric air and exposed to the light of the sun ; but, if these facts do not affect the subject of our enquiry, we may pass them over as if they had no existence. When, therefore, we employ the expression 'only one circumstance in common,' we must be understood to mean 'only one *material* circumstance,' and to exclude all circumstances which a wide experience or previous inductions have shown to be *immaterial* to the question before us. It need hardly be added that, in forming this judgment as to the material or immaterial character of the circumstances, the greatest caution is often required.

But, suppose we have ascertained (when enquiring into the cause of a given effect) that the instances agree in only one antecedent (or rather one *material* antecedent), namely A, and that we have so multiplied and varied the instances as to have satisfied ourselves that we have excluded the possibility of a Plurality of Causes, are we justified in drawing the inference that A is the cause of *a*? We are so justified, for *a* must be due to something which went before it, and, as it has been shown that it is not due to any of the other antecedents, it must be due to A. Similarly, if our object be to enquire into the effect of a given cause A, we are justified, if we discover a consequent *a*, of which we can assure ourselves that it is not due to any of the other antecedents, in regarding it as the effect of A.

Hitherto, we have supposed the antecedents and consequents to be discriminated. But, suppose that we have a number of phenomena a b c d e, a d e f g, &c., in which we cannot discriminate them, how will the conclusions of the Method of Agreement be affected? There will, as in the former cases, obviously be the difficulties arising from Plurality of Causes and the complexity of the phenomena. Supposing, however, these to be overcome, and two circumstances only, a and b, to have been ascertained to be common to all the instances, what conclusion shall we be justified in drawing with reference to the connexion between a and b? It is only reasonable to suppose that they must be causally connected in some way, else their connexion would be a mere casual coincidence; a supposition which we assume to have been excluded by the number and variety of the in-

stances examined.   But they need not necessarily stand
to each other in the relation of cause and effect, for they
may be common effects (in the same, or in different degrees of descent) of some cause which has itself ceased
to operate.   In social and physiological phenomena this
is frequently the case.   A disease will leave effects behind
it which will continue to co-exist for years after the disease
itself has passed away, and which, though not standing
to each other in the relation of cause and effect, are thus
causally connected.   The social condition of any old
country is, to a great extent, an aggregate of such effects,
the original cause or causes of which have long ceased
to have any existence.

It should be noticed that the Method of Agreement is
mainly, though not exclusively, a Method of Observation
rather than of Experiment, and that it is applied far more
frequently for the purpose of enquiring into the causes of
given effects than into the effects of given causes.   The
reason of this peculiarity is that in trying an experiment,
or in enquiring into the effect of a given cause, we are
generally able to employ one of the other Methods,
which, as will be seen hereafter, are not exposed to the
same difficulties as the Method of Agreement.

It should also be noticed that where, after a careful
elimination and an examination of a sufficiently large
number of instances, we have, instead of two, some three,
four, or more circumstances common to all the instances,
we may, with much probability, regard them all, unless
we know or suspect any of them to be immaterial cir-

cumstances, as being causally connected. If the common circumstances be a, b, c, d, this is all that we can infer. But, if they be ,A, B, C, a, we may infer that the cause of a is certainly either A or B or C, or some two of them acting jointly, or all acting together, while those common antecedents, which do not either constitute or contribute to the cause, probably stand in some causal relation to it, and consequently to its effect a. Similar conclusions may be drawn, if the common circumstances left after elimination be A, a, β, γ. Thus, for instance, a, β, γ might all be joint effects of A, or a might be its immediate effect, and β, γ effects of a, and so on.

It is perhaps not superfluous to remind the student that, in the application of this Method, he should be peculiarly careful not to overlook any instance in which the given phenomenon is unaccompanied by the other circumstance. Such an instance should at once lead him to suspect that some third common circumstance, which may be the true cause (or effect) of the given phenomenon, has escaped his attention, but this, if it be the case, does not necessarily vitiate his conclusion. If the given phenomenon be the consequent, and this other circumstance the antecedent, such an instance may only point to some other and independent cause of the phenomenon in addition to the cause he supposes himself to have ascertained. If, on the other hand, the given phenomenon be the antecedent, and this other circumstance the consequent, such an instance may only point to a counteracting cause which, in this exceptional case, frustrates the supposed effect. The only condition essential to an application of the Method of Agreement is that the cases on which the inference is founded shall present only two circumstances in common. It is not necessary that these circumstances should invariably be found in conjunction, provided that in the cases where they are found in conjunction no other common circumstance can be detected. I shall recur to this subject below [8].

In the statement of the Canon, I have thought it desirable to introduce the expression ' with more or less of probability,' in order to show that, under no circumstances, does an inference drawn in accordance with the Method of Agreement attain to absolute and formal certainty, though, as we have seen, it may attain to moral certainty.

As familiar examples of the employment of the Method of Agreement, the following may be adduced :—

After taking a particular kind of food, whatever else I may eat or drink, and however various my general state of health, the temperature of the air, the climate in which I am living, and my divers other surroundings, I am invariably ill; I am justified in regarding the food as the probable cause of my illness, and avoid it accordingly. This example furnishes a good illustration both of the difficulties and of the possible cogency of the Method of Agreement. What made me ill on each of two, three, or four occasions, may have been some viand different from the

[8] See pp. 145-7.

one in question, but it is very unlikely, if the number of occasions on which the inference is based be considerable, that it has been a different viand on each of them.

I find that a certain plant always grows luxuriantly on a particular kind of soil; if my experience of the other conditions be sufficiently various, I am justified in concluding that the soil probably possesses certain chemical constituents which are peculiarly favourable to the production of the plant.

Trade is observed to be in a languishing condition wherever there exist certain restrictions, such as high duties, difficulties thrown in the way of landing or locomotion, &c.; if it could be ascertained that these countries agreed in no other respect which could influence the condition of trade, except in being subject to these restrictions, it might be inferred with considerable probability that the commercial depression was due to the restrictions as a cause.

In all these cases, it will be seen that the great difficulty consists in ascertaining that the supposed cause is the only circumstance, or the only material circumstance, which, in addition to the phenomenon itself, the various instances possess in common.

I now append a few instances of a less familiar nature:—

The occurrence of Aurora Borealis has, under meteorological conditions of very different character, been invariably found to be accompanied by considerable magnetic disturbances. It is rightly inferred that there is some causal connexion between magnetic disturbance and the occurrence of the Aurora Borealis.

It has been observed uniformly, and under a variety of circumstances, that, wherever an indiscriminate system of almsgiving has prevailed, the population has, sooner or later, become indolent and pauperised. This fact may be noticed especially in the neighbourhood of large monasteries, in parishes where large sums of money are distributed in the shape of 'doles,' in places which are the residence of rich and charitable but injudicious persons, and the like. The reason is not difficult to discover. The unfortunate recipients of the charity are left without the ordinary motives to exertion, and consequently, when the abnormal supply ceases, or becomes too small for the wants of an increased population, being without self-reliance or any special skill, they have no resource but beggary.

After a variety of experiments on substances of the most different kinds, and under the most different circumstances, it has been found that, as a body passes from a lower degree of temperature to a higher, it invariably undergoes a change of volume, though that change may not always be in the same direction, it being, in the great majority of cases, in the direction of expansion, but, occasionally, in that of contraction. Hence it has been inferred that change of volume is an invariable effect of change of temperature (it being understood, of course, that pressure and other circumstances, as, for instance, the chemical condition of the body, remain the same).

It has been supposed by some writers on physics that we may go further than this conclusion, and state that augmentation of temperature is invariably followed by augmentation of volume, and diminution of temperature by diminution of volume, the exceptions of water[9] as well as of bismuth and of the casting-metals generally (which suddenly expand at the moment of solidification) being explained as anomalies due to some interfering cause. We are, however, at present so little acquainted with the intimate constitution of bodies, that it might be rash to state the proposition in this form, and, stated as above, it is open to no exception[10].

[9] Water follows the general rule, and continues to contract in bulk as its temperature is lowered, till it reaches about 39° Fahrenheit or 4° Centigrade, when it begins to expand, and continues to do so till after its conversion into ice, so that a given weight of water at the temperature (say) of 37°, or when frozen, occupies more space than it occupied at (say) the temperature of 40°. This anomaly is somewhat boldly explained by Sir W. Grove as due to the setting in of the process of crystallization, which he supposes to begin at 39°, and to interfere with the ordinary law of contraction and expansion. (See Grove's *Correlation of Physical Forces*, fifth ed. p. 58, &c.)

[10] I adduce this instance as an example of the Method of Agreement rather than of the Method of Concomitant Variations, because the argument, as here stated, rests rather upon the variation of circumstances and the great diversity of bodies in which the law is found to hold good, than upon the relation between the various degrees of expansion or contraction and the various degrees of temperature in the same body. Had the stress been laid upon the latter consideration, the argument would undoubtedly have been an instance of the Method of Concomitant Variations.

It frequently happens, in fact, that two or more Methods are combined in the same proof. In the present instance, as will be seen below, the argument as applied to each particular kind of body

The following example, which also illustrates the caution necessary to be observed in framing a general proposition, is extracted from Sir John Herschel's *Discourse on the Study of Natural Philosophy* [11] :—

'A great number of transparent substances, when exposed in a certain particular manner, to a beam of light which has been prepared by undergoing certain reflexions or refractions (and has thereby acquired peculiar properties, and is said to be "*polarized*"), exhibit very vivid and beautiful colours, disposed in streaks, bands, &c. of great regularity, which seem to arise within the substance, and which from a certain regular succession observed in their appearance, are called "periodical colours." Among the substances which exhibit these periodical colours occur a great variety of transparent solids, but no fluids and no opaque solids. Here, then, there seems to be sufficient community of nature to enable us to use a general term, and to state the proposition as a law, viz. *transparent solids* exhibit periodical colours by exposure to polarized light. However, this, though true of many, does not apply to *all* transparent solids, and therefore we cannot state it as a general truth or law of nature in this form; although the reverse proposition, that all solids which exhibit

(mercury, for instance) is an argument based on the Method of Concomitant Variations; but when we proceed to extend the experiment to other bodies, and then argue from the variety of the bodies examined that a body, in passing from one degree of temperature to another, invariably undergoes a change of volume, it appears to me that we are no longer employing the Method of Concomitant Variations but the Method of Agreement. It must be borne in mind that the object of our enquiry is not strictly the effects of heat (for the total effects of heat, inasmuch as we cannot wholly exhaust any body of its heat, must be unknown to us), but the effects of a change of temperature.

[11] § 90.

such colours in such circumstances are *transparent*, would be correct and general. It becomes necessary, then, to make a list of those to which it does apply; and thus a great number of substances of all kinds become grouped together in a class linked by this common property. If we examine the individuals of this group, we find among them the utmost variety of colour, texture, weight, hardness, form, and composition; so that, in these respects, we seem to have fallen upon an assemblage of contraries. But, when we come to examine them closely in all their properties, we find they have all one point of agreement, in the property of double refraction, and therefore we may describe them all truly as *doubly refracting substances*. We may, therefore, state the fact in the form, "Doubly refracting substances exhibit periodical colours by exposure to polarized light;" and in this form it is found, on further examination, to be true, not only for those particular instances which we had in view when we first propounded it, but in all cases which have since occurred on further enquiry, without a single exception; so that the proposition is general, and entitled to be regarded as a law of nature.'

The experiments by which Dr. Wells [12] established his Theory of Dew afford a remarkable example of the Method of Agreement. By employing various objects of different material under a variety of circumstances,

---

[12] Dr. Wells' *Memoir on the Theory of Dew*, which had become very scarce, was reprinted by Longmans and Co. in 1866. It is very brief, and well deserves to be carefully read by every student of scientific method. Sir John Herschel (*Natural Philosophy*, § 168) speaks of the speculation as 'one of the most beautiful specimens' he can call to mind ' of inductive experimental enquiry lying within a moderate compass.' Mr. Mill also employs it as one of his Miscellaneous Examples in Bk. III. ch. ix. of his *Logic*.

he showed that, whatever the texture of the object, the state of the atmosphere, &c., it is an invariable condition of the deposition of dew that the object on which it is deposited shall be colder than the surrounding atmosphere, the greater coldness of the object being itself produced by the radiation of heat from its surface. This, to quote the words of Sir John Herschel, is the case not only with 'nocturnal dew,' but with 'the analogous phenomena' of 'the moisture which bedews a cold metal or stone when we breathe upon it; that which appears on a glass of water fresh from the well in hot weather; that which appears on the inside of windows when sudden rain or hail chills the external air; that which runs down our walls when, after a long frost, a warm moist thaw comes on.'

It is by the Method of Agreement that we discover the symptoms of a disease, the signs of a political revolution, national characteristics, the order of superposition among geological strata, grammatical rules, and the like.

The first division of Bacon's *instantiæ solitariæ* coincides with the cases contemplated in the Method of Agreement, as the second coincides with the cases contemplated in the Method of Difference. The example employed in the first is so remarkable both in itself, and as an anticipation of Newton's Speculations on Colour, that I may adduce it as an additional instance of the Method of Agreement :—

'Exempli gratia: si fiat inquisitio de natura *coloris*, *instantiæ solitariæ* sunt prismata, gemmæ crystallinæ, quæ reddunt colores, non solum in se, sed exterius supra parietem. Item rores, &c. Istæ enim nil habent commune cum coloribus fixis in floribus, gemmis coloratis, metallis, lignis, &c. præter ipsum colorem. Unde facile colligitur, quod color nil aliud sit quam modificatio imaginis lucis immissæ et receptæ: in priore genere, per gradus diversos incidentiæ; in posteriore, per texturas et schematismos varios corporis. Istæ autem *instantiæ* sunt *solitariæ* quatenus ad similitudinem [13].'

In attempting to ascertain the cause of a given effect, *a*, it may happen that we find a particular antecedent, A, frequently, but not invariably, accompanying it. If, in those cases which present both *a* and A, no other common circumstance can be detected, we may infer  that A is probably a cause of *a*. I say 'a cause,' for the fact that *a* may be present without A is a proof that A is not the only cause. My meaning will be plain from the following example :—

We compare instances in which bodies are known to assume a crystalline structure, but which have no other point of agreement; in the great majority of instances, though not in all, we find that these bodies have assumed their crystalline structure during the process of solidification from a fluid state, either gaseous or liquid, and, so far as we can ascertain, these cases have no other circumstance in common. From these facts it may be reasonably inferred that the passage from a fluid to

[13] *Novum Organum*, Lib. II. aph. xxii.

a solid state is a cause, though not the only cause, of crystallization [14].

Again, when A is frequently, though not invariably, followed by *a*, and there is, so far as we can ascertain, no other common antecedent, we are justified in suspecting that A is a cause of *a*, and that, in the cases where *a* does not occur, the operation of A is counteracted by some other cause. If, for example, a certain occupation or mode of living is found to be usually, though not invariably, attended by a particular form of disease, we seem to be justified in regarding this occupation or mode of living as a cause of the disease, and in explaining the few cases in which the disease does not occur as due to exceptional and counteracting circumstances.

Similarly, when a and b are found in frequent, though not invariable, conjunction [15], and, in the cases where

[14] This example is adopted, with considerable modifications, from one which occurs in Mr. Mill's *Logic*, Bk. III. ch. viii. § 1. I am indebted to Sir John Herschel for pointing out to me that Mr. Mill's example (which I had originally adopted as it stood) is too broadly stated. 'The solidification of a substance from a liquid [it should be fluid] state' is not 'an invariable,' but only an usual 'antecedent of its crystallization.' The reader will find several exceptions noticed in Watts' *Dictionary of Chemistry*, art. Crystallization.

[15] The invariable conjunction of two phenomena, when the presence of the one implies the presence of the other, and the absence of the one the absence of the other, is a case falling under the Double Method of Agreement, to be explained presently; but those cases, in which we simply know that a given phenomenon is invariably preceded or invariably followed by another, fall under the Method of Agreement just discussed. If a given phenomenon is, so far as we know, invariably preceded by another, this fact justifies

they are found together, there occurs, so far as we can ascertain, no other common circumstance, we are justified in suspecting that there exists some causal connexion between them.

The student, who is acquainted with the science of Medicine, will find a good illustration of the extreme difficulty attending the application of the Method of Agreement, as well as of the Joint Method of Agreement and Difference (to be noticed presently), in the disputes which still occur as to the cause of the mental disease which is known as Atactic Aphasia, that is, the condition in which, with reference to certain sounds, the patient has lost the power of co-ordinating the muscles of speech. The French physiologist, M. Broca, laid down the position that this disease is invariably due to a lesion of the third frontal convolution of the left hemisphere of the brain, the disease being invariably attended by the specific lesion, and the lesion never occurring without the disease. His followers maintain that the instances are decisive in favour of this theory, while the apparent exceptions admit

us in suspecting (though it does not prove) that the antecedent is not only *a* cause, but the *only* cause, of the given phenomenon. Such a conclusion can only be *proved* (even approximately) by the Double Method of Agreement. It is, however, as already pointed out, not in the invariableness of the conjunction, but in the fact that the instances examined present, so far as we can ascertain, only two phenomena in common, that the cogency of the Method of Agreement consists. But of this fact invariableness of antecedence (or of consequence) furnishes one of the strongest proofs, inasmuch as such invariableness implies a very wide variation of circumstances; hence the stress laid upon it in some of the examples adduced above.

of a satisfactory explanation; his opponents, on the other hand, assert that there are well-established cases, both of atactic aphasia without the specific lesion, and of the lesion without aphasia [16].

METHOD OF DIFFERENCE.

CANON.

If an instance in which the phenomenon under investigation occurs, and an instance in which it does not occur, have every circumstance in common save one, that one occurring only in the former; the circumstance in which alone the two instances differ, is the effect, or cause, or a necessary part of the cause, of the phenomenon.

The circumstances a, b, c are found in conjunction with d, e, f, and the omission or disappearance of the circumstance a is found to be attended by the disappearance of the circumstance d. It is inferred that a and d are so connected that one is cause (or a necessary part of the cause) and the other effect. If, moreover, it can be ascertained that a is the antecedent and d the con-

[16] See a paper by Dr. William Ogle in the *St. George's Hospital Reports*, vol. ii.; a Pamphlet by Dr. Frederic Bateman of Norwich, published by J. E. Adlard, Bartholomew Close, London, 1868; Dr. Reynolds' *System of Medicine*, vol. ii. pp. 442–444; and various reports of discussions published in the *Lancet* and other medical journals. I have to thank my friends, Professors Acland and Rolleston, for their kindness in supplying me with information on this interesting subject, and regret that my space prevents me from pursuing it at greater length.

sequent, or that, though there are instances in which d occurs without a, there are no instances in which a occurs without d, we may proceed to infer (in the latter case, on the ground that a phenomenon may have more than one cause, but that a cause, unless counteracted by some other cause, must be attended by its effect) that a is the cause, and d the effect. Similarly, if the circumstances a, b, c are found in conjunction with d, e, f, and the introduction of the circumstance x into the former set of phenomena is found to be attended by the appearance of the circumstance y in the latter set of phenomena (so that they may be represented respectively as a, b, c, x; d, e, f, y), it may be inferred that x and y are related as cause and effect; or, if x be the antecedent and y the consequent, or the appearance of x be invariably attended by the appearance of y while the appearance of y is not invariably attended by the appearance of x, that x is the cause and y the effect. The reasons on which the Canon rests are obvious. All other circumstances remaining the same, if the introduction or omission of any circumstance be followed by a change in the remaining circumstances, that change must be due to such introduction or omission, as an effect to a cause; or, if two new circumstances enter simultaneously, without producing any other change in the phenomenon, these two circumstances (except on the improbable supposition that they are two causes exactly counteracting each other) must be related as cause and effect, though we may be unable to say which of the two is cause and

which effect. 'The Method of Agreement,' says Mr. Mill, 'stands on the ground that whatever can be eliminated, is not connected with the phenomenon by any law. The Method of Difference has for its foundation that whatever can *not* be eliminated, *is* connected with the phenomenon by a law.' In the Method of Difference, the instances agree in everything, except in the possession of two circumstances which are present in the one instance and absent in the other. In the Method of Agreement, the various instances compared (for here we generally require more than two instances) agree in nothing, except in the possession of two circumstances which are common to all the instances. One Method is called the Method of Agreement, because we compare various instances to see in what they agree; the other is called the Method of Difference, because we compare an instance in which the phenomenon occurs with another in which it does not occur, in order to see in what they differ.

Instances of the Method of Difference are not far to seek. A piece of paper is thrown into a stove; we have no hesitation in regarding its apparent consumption as the effect of the heat of the fire, for we feel assured that the sudden increase of temperature is the only new circumstance to which the piece of paper is exposed, and that, therefore, any change in the condition of the paper must be due to that cause. A bullet is fired from a gun, or a dose of prussic acid is administered, and an animal instantly falls down dead. There is no hesitation in ascribing the death to the gun-shot wound or the dose of poison. Nor is this confidence the effect of any wide experience, for, if it were the first time that we had seen a gun fired or a dose of poison administered, we should have no hesitation in ascribing the altered condition of the animal to this novel cause; we should know that there was only one new circumstance operating upon it, and, consequently, that any change in its condition must be due to that one circumstance. In all these instances, there is the introduction of a new antecedent, x, to which the new consequent, y, must be due. But, if the omission of one circumstance be attended by the omission of another, we may argue with equal confidence. I withdraw my hand from this book which is resting upon it, and the book instantly falls to the ground; there is no hesitation in referring the altered position of the book to the withdrawal of my support. A man is deprived of food, and he dies; we have no hesitation in ascribing the disappearance of the phenomenon we call life to the withdrawal of the means by which it is maintained. In these instances, we have certain antecedents, followed by certain consequents, and, observing the simultaneous or successive disappearance of A and a, we have no hesitation in connecting the two as cause and effect.

All *crucial instances* (instantiæ [17] crucis, as they are

[17] 'Inter prærogativas instantiarum ponemus loco decimo quarto *instantias crucis;* translato vocabulo a *crucibus,* quæ, erectæ in biviis, indicant et signant viarum separationes. Has etiam *instantias*

called by Bacon) are applications of the Method of Difference. A crucial instance is some observation or experiment which enables us at once to decide between two or more rival hypotheses. It will be familiar to every one in the form of the *chemical test*, as where we apply an acid for the purpose of determining the character of a metal, or a metal for the purpose of detecting latent poison. According to the metaphor, there are two or more ways before us, and the observation or experiment acts as a 'guide-post' (crux) in determining us which to take. The following beautiful example of a Crucial Instance is borrowed from Sir John Herschel[18].

'A curious example is given by M. Fresnel, as decisive, in his mind, of the question between the two great opinions on the nature of light, which, since the time of

decisorias, et *judiciales*, et in casibus nonnullis *instantias oraculi*, et *mandati*, appellare consuevimus. Earum ratio talis est. Cum in inquisitione naturæ alicujus, intellectus ponitur tanquam in æquilibrio, ut incertus sit, utri naturarum e duabus, vel quandoque pluribus, causa naturæ inqusitæ attribui aut assignari debeat, propter complurium naturarum concursum frequentem et ordinarium ; *instantiæ crucis* ostendunt consortium unius ex naturis (quoad naturam inquisitam) fidum et indissolubile, alterius autem varium et separabile ; unde terminatur quæstio, et recipitur natura illa prior pro causa, missa altera et repudiata. Itaque hujusmodi instantiæ sunt maximæ lucis, et quasi magnæ auctoritatis ; ita ut curriculum interpretationis quandoque in illas desinat, et per illas perficiatur. Interdum autem *instantiæ crucis* illæ occurrunt et inveniuntur inter jampridem notatas ; at ut plurimum novæ sunt, et de industria atque ex composito quæsitæ et applicatæ, et diligentia sedula et acri tandem erutæ.'—*Novum Organum*, Lib. II. aph. xxxvi.

[18] *Discourse on the Study of Natural Philosophy*, § 218.

Newton and Huyghens, have divided philosophers;'—
that is, between what is called 'the emission theory,'
according to which light consists of actual particles
emitted from luminous bodies, and what is called 'the
undulatory theory,' according to which light consists in
the vibrations of an elastic medium pervading all space.

'When two very clean glasses are laid one on the other,
if they be not perfectly flat, but one or both in an almost imperceptible degree convex or prominent, beautiful and vivid
colours will be seen between them ; and if these be viewed
through a red glass, their appearance will be that of alternate
dark and bright stripes. These stripes are formed *between*
the two surfaces in apparent contact, as any one may satisfy
himself by using, instead of a flat *plate* of glass for the upper
one, a triangular-shaped piece, called a prism, like a three-
cornered stick, and looking through the inclined side of it
next the eye, by which arrangement the reflexion of light
from the upper surface is prevented from intermixing with
that from the surfaces in contact. Now, the coloured stripes
thus produced are explicable on both theories, and are appealed
to by both as strong confirmatory facts ; but there is a difference in one circumstance according as one or the other
theory is employed to explain them. In the case of the
Huyghenian doctrine, the intervals between the bright stripes
ought to appear *absolutely black;* in the other, *half bright*,
when so viewed through a prism. This curious case of difference was tried as soon as the opposing consequences of
the two theories were noted by M. Fresnel, and the result is stated by him to be decisive in favour of that theory
which makes light to consist in the vibrations of an elastic
medium[19].

[19] Mr. Mill (*Logic*, Bk. III. ch. xiv. § 6) maintains that it does not
follow from this experiment that 'the phenomena of light are results

The following is an example of a similar kind. It had been determined, from theoretical considerations, that, on the assumption of the undulatory theory, the velocity of light must be less in the more highly refracting medium, while, according to the emission theory, it ought to be greater. When M. Foucault had invented his apparatus for determining the velocity of light, it became possible to submit the question to direct experiment; and it was established by himself and M. Fizeau that the velocity of light is less in water (the more highly refracting medium) than in air, in the inverse proportion of the refractive indices. The result is, therefore, decisive in favour of the undulatory, or, at least, against the emission theory [20].

There is no science, perhaps, in which the Method of Difference is so extensively used as the science of Chemistry, and that because chemistry is emphatically a science of experiment. Almost any chemical experiment will serve as an instance of the Method of Difference. Mix, for example, chloride of mercury with iodide of potassium, and the result will be a colourless liquid at the top of the vessel with a brilliant red precipitate at

of the laws of elastic fluids, but at most that they are governed by laws partially identical with these.' But, though the experiment may not be decisive as in favour of the Undulatory Theory, it is undoubtedly decisive as against the Emission Theory. It may be necessary to add that the term 'fluids' would now be repudiated by those who hold the Undulatory Theory.

[20] See Lloyd's *Wave Theory of Light,* 3rd ed. Arts. 41, 42 ; Ganot's *Physics,* English translation, 12th edition, Art. 506.

the bottom. There can be no hesitation in ascribing this result to the mixture of the two liquids; and two similar experiments will enable us to determine that the chlorine has been set free from the mercury and united with the potassium, which itself has been set free from the iodine with which it was previously united, while the iodine has united with the mercury, the former producing chloride of potassium (dissolved in the colourless liquid), the latter iodide of mercury (the red precipitate).

The science of Heat (or, as Dr. Whewell proposes to call it, Thermotics) also furnishes excellent examples of the Method of Difference. The following instances are adapted from Professor Tyndall's *Heat a Mode of Motion*[21] :—

'Here is a brass tube, four inches long, and of three-quarters of an inch interior diameter. It is stopped at the bottom and screwed on to a whirling table, by means of which the upright tube can be caused to rotate very rapidly. These two pieces of oak are united by a hinge, in which are two semicircular grooves, intended to embrace the brass tube. Thus the pieces of wood form a kind of tongs, the gentle squeezing of which produces friction when the tube rotates. I partially fill the tube with cold water, stop it with a cork to prevent the splashing out of the liquid, and now put the machine in motion. As the action continues, the temperature of the water rises, and now the tube is too hot to be held in the fingers. Continuing the action a little longer, the cork is driven out with explosive violence, the steam which follows it producing by its precipitation a small cloud in the atmosphere.'

[21] Third Edition, ch. i. §§ 14–16.

In this experiment it will be noticed that only one new antecedent is introduced, namely the motion of the machine; hence the increased temperature of the water and the various effects which follow upon it are due to the motion as a cause. The experiment, then, shows that heat is generated by the action of mechanical force.

The converse of this proposition, namely that heat is consumed in mechanical work, or, as it is often stated, transmuted into mechanical energy, is proved by the two next experiments.

'This strong vessel is filled at the present moment with compressed air. It has lain here for some hours, so that the temperature of the air within the vessel is now the same as that of the air of the room without it. At the present moment this inner air is pressing against the sides of the vessel, and if this cock be opened a portion of the air will rush violently out. The word "rush," however, but vaguely expresses the true state of things; the air which issues is driven out by the air behind it; this latter accomplishes the work of urging forward the stream of air. And what will be the condition of the *working air* during this process? It will be chilled. The air executes work, and the only agent it can call upon to perform the work is the heat to which the elastic force with which it presses against the sides of the vessel is entirely due. A portion of this heat will be consumed, and a lowering of temperature will be the consequence. Observe the experiment. I will turn the cock, and allow the current of air from the vessel to strike against the face of the pile[22].

[22] That is, the thermo-electric pile, a delicate instrument for indicating very small changes of temperature. It is by means of this instrument that it has recently been shown that we receive heat (though, of course, in infinitesimal quantities) from the moon's rays.

The magnetic needle instantly responds ; its red end is driven towards me, thus declaring that the pile has been *chilled* by the current of air.'

' Here moreover is a bottle of soda-water, slightly warmer than the pile, as you see by the deflexion it produces. Cut the string which holds it, the cork is driven out by the elastic force of the carbonic acid gas ; the gas performs work, in so doing it consumes heat, and now the deflexion produced by the bottle is that of cold.'

The last experiment furnishes a good instance of the extreme simplicity of the examples by which scientific truths may often be illustrated.

The uncertainty which, as we have seen, always attaches to conclusions arrived at by the Method of Agreement renders it desirable that they should, wherever it is possible, be confirmed by an application of the Method of Difference. A beautiful instance of such a confirmation is adduced by Mr. Mill in the case of Crystallization, The Method of Agreement has already led us to the conclusion that the solidification of a substance from a fluid state is a very frequent antecedent of its crystallization, and so, probably, one, at least, of its causes. But the Method of Difference completes the evidence, and enables us to state positively that it is a cause.

' For in this case we are able, after detecting the antecedent A, to produce it artificially, and, by finding that *a* follows it, verify the result of our induction. The importance of thus reversing the proof was never more strikingly manifested than when, by keeping a phial of water charged with siliceous particles undisturbed for years, a chemist (I believe Dr. Wollaston) succeeded in obtaining crystals of quartz; and in the equally interesting experiment in which Sir James Hall produced artificial marble, by the cooling of its materials from fusion under immense pressure : two admirable examples of the light which may be thrown upon the most secret processes of nature by well-contrived interrogation of her [23].'

It will be noticed that the Method of Difference is specially adapted to the discovery of the effects of given causes, whereas, where it is our object to discover the cause of a given effect, we are usually compelled to have recourse to the Method of Agreement. The Method of Agreement is, in fact, mainly a Method of Observation, whereas the Method of Difference is mainly a Method of Experiment. We may indeed arrange the conditions of an experiment so as to satisfy the requirements of the Method of Agreement, and Nature may (as in the familiar case of lightning) herself satisfy the requirements of the Method of Difference, but, as a rule, it will be found that arguments based on observations fall under the former, and arguments based on experiments under the latter Method. It is hardly necessary to add that, wherever we have our choice between the two methods, we should invariably select the Method of Difference.

[23] Mill's *Logic,* Bk. III. ch. viii. § 1. I have been obliged, in accordance with what has been said on p. 146, to state, with considerable modifications, the conclusion in this instance as arrived at by the Method of Agreement. The account of the application to it of the Method of Difference has been stated in Mr. Mill's own words.

In the employment of the Method of Difference, the greatest care should be taken to introduce only one new antecedent, or at least only one new antecedent which can influence the result. As the whole force of the argument based on this Method depends on the assumption that any change which takes place in the phenomenon is due to the antecedent then and there introduced, it is plain that we can place no reliance on our conclusion unless we feel perfectly assured that no other antecedent has intervened. If, for instance, it is our object to ascertain the temperature of the atmosphere, we must take the greatest care that our thermometer is not affected by the heat radiated from or conducted by other bodies. The most curious examples of the disregard of this caution may be found in the History of Medicine. Something perfectly inert has been administered to a patient in combination with some powerful drug, some important alterations in his diet, or some strict régime; then the effects of the drug, diet, or régime have been unwittingly ascribed to the inert substance. Had the ancients recognised that instead of one cause acting on falling bodies, as appeared to them to be the case, there were really two, the action of gravity tending downwards and the resistance of the atmosphere pressing upwards, they could never have fallen into the gross error of supposing that bodies fall in times inversely proportional to their weights.

DOUBLE METHOD OF AGREEMENT.

CANON.

If two or more instances in which the phenomenon occurs have only one other circumstance in common, while two or more instances, falling within the same department of investigation[24], from which the phenomenon is absent have nothing in common save the absence of that circumstance; that circumstance is the effect, or the cause, or a necessary part of the cause, of the phenomenon. Moreover (supposing the requirements of the Method to be rigorously fulfilled), the circumstance proved by the Method to be the cause is the only cause of the phenomenon.

The uncertainty attaching to the Method of Agreement may, even where it is impossible to have recourse to the Method of Difference, be, to some extent, remedied by the employment of what is called by Mr. Mill the Joint Method of Agreement and Difference, or the Indirect Method of

---

[24] In recent editions I have inserted in the statement of the Canon the words 'falling within the same department of investigation,' because, as has been pointed out to me, the student might otherwise not see that, for the purposes of comparison, the positive and negative instances must be *in pari materiâ.* Thus, if the subject of enquiry is language, the negative as well as the positive instances must be sought in the department of language; or, if the subject of enquiry lies within the sphere of morals, or of physical forces, or of living organisms, the negative as well as the positive instances must be sought within those respective departments. Practically, however, there is no occasion for definite rules on this head, as the common-sense of the investigator is quite sufficient to keep him within the limits of the enquiry.

Difference. This Method consists in a double employment of the Method of Agreement and a comparison of
the results thus obtained, the comparison assimilating it
to the Method of Difference. We, first of all, compare
cases in which the phenomenon occurs, and, so far as we
can ascertain, find them to agree in the possession of
only one other circumstance. But, though we may not
be justified in regarding this inference as certain, we may
increase our assurance by proceeding to compare cases
in which the phenomenon does not occur. If these cases
agree in nothing but the non-possession of the circumstance which the other cases agreed in possessing, we
have a set of negative instances agreeing in nothing but
the absence of the given phenomenon and the absence
of the aforesaid circumstance. The set of negative
instances may now be compared with the set of positive
instances, and we may argue thus: The positive instances agree in nothing but the presence of the given
phenomenon and this other circumstance, and the negative instances agree in nothing but the absence of the
given phenomenon and this other circumstance. Hence
we may regard it as a highly probable inference that
they are connected together as cause and effect. I
say 'highly probable,' for, as we are not absolutely
certain that the conditions of the Method of Agreement
have been satisfied in the case of the positive instances,
so, from the extreme difficulty of proving a negative,
we must be still less certain that they have been satisfied
in the case of the negative instances. What (in addition

M

to another advantage, to be noticed presently) is gained by the Method is a sort of double assurance, so far as the assurance goes. *If* the one set of instances agreed in nothing but the presence of the two circumstances, and *if* the other set of instances agreed in nothing but the absence of the two circumstances, then we should be able to infer, by the Method of Difference, that the introduction of the given phenomenon (which we will suppose to be the consequent) always follows on the introduction of the other circumstance (which we will suppose to be the antecedent), and, *vice versâ,* that the removal of the given phenomenon always follows on the removal of the other circumstance, or, in other words, that the given phenomenon is the *effect* and the other circumstance the *cause.*

But this Method, supposing its conditions to be rigorously satisfied, possesses one advantage peculiar to itself. The Single Method of Agreement, as we have seen, is always theoretically open to the objection arising from Plurality of Causes, but this Method, if the set of  negative instances be perfect, is not only free from that objection, but also sustains the conclusion that the inferred cause is the *only* cause of the phenomenon in question (or, in case we do not know which is antecedent and which is consequent, that a and b are so connected that one of them is the cause and the only cause of the other). ' In the joint method,' says Mr. Mill[25], ' it is supposed not only that the instances in

[25] Mill's *Logic*, Bk. III. ch. x. § 2.

which a is agree only in containing A, but also that the instances in which a is not agree only in not containing A. Now, if this be so, A must be not only the cause of a, but the only possible cause: for if there were another, as for example B, then in the instances in which a is not, B must have been absent as well as A, and it would not be true that these instances agree *only* in not containing A.' It may be asked, then, if the negative branch of the argument be so forcible, why should we employ the positive branch? It is by means of the positive branch that we are, as it were, put on the track of the one other circumstance in which the instances presenting the given phenomenon agree, and by means of the negative branch that we prove the accuracy of our conclusion. 'It is generally,' continues Mr. Mill, 'altogether impossible to work the Method of Agreement by negative instances without positive ones: it is so much more difficult to exhaust the field of negation than that of affirmation.'

It is plain that the conditions of the Joint Method can only be rigorously fulfilled where there is an invariable conjunction between two phenomena, so that the two are (unless counteracting circumstances intervene) always present together and always absent together. For, if A be the *only* cause of a, the effect a cannot be present without the cause A, nor can the cause A be present without being attended by the effect a. Hence, invariable conjunction may be regarded as a sign that the conditions of this Method are fulfilled, and it is from the observation of such an invariable conjunction that the argument frequently proceeds. In such cases, the number of instances, both positive and negative, which have been observed, is supposed to be so great and of such variety as to have excluded all other common circumstances except the presence or absence of the two phenomena in question.

The Joint Method of Agreement and Difference (or the Indirect Method of Difference, or, as I should prefer to call it, the Double Method of Agreement) is being continually employed by us in the ordinary affairs of life. If, when I take a particular kind of food, I find that I invariably suffer from some particular form of illness, whereas, when I leave it off, I cease to suffer, I entertain a double assurance that the food is the cause of my illness. I have observed that a certain plant is invariably plentiful on a particular soil; if, with a wide experience, I fail to find it growing on any other soil, I feel confirmed in my belief that there is in this particular soil some chemical constituent, or some peculiar combination of chemical constituents, which is highly favourable, if not essential, to the growth of the plant.

Dr. Wells' *Essay on the Theory of Dew* presents an extremely instructive instance of the application of the Double Method of Agreement :—

'It appears' (I am here quoting from Mr. Mill[26]) 'that the instances in which much dew is deposited, which are very various, agree in this, and, so far as we are able to observe, in this only, that they either radiate heat rapidly or conduct

[26] Mill's *Logic*, Bk. III. ch. ix. § 3.

it slowly : qualities between which there is no other circumstance of agreement than that, by virtue of either, the body tends to lose heat from the surface more rapidly than it can be restored from within. The instances, on the contrary, in which no dew, or but a small quantity of it, is formed, and which are also extremely various, agree (as far as we can observe) in nothing except in *not* having this same property. We seem, therefore, to have detected the characteristic difference between the substances on which dew is produced, and those on which it is not produced. And thus have been realized the requisitions of what we have termed the Indirect Method of Difference, or the Joint Method of Agreement and Difference.'

Several beautiful illustrations of the Joint Method of Agreement and Difference may be found in the recent discoveries made by means of Spectrum Analysis. I shall select one which is peculiarly interesting on account of its employment in the attempt to determine the constitution of the sun and some of the other heavenly bodies.

A very thin sheet of light proceeding from incandescent hydrogen is passed through a prism, and it is invariably found (with the exception of the third case mentioned in note 27) that in the spectrum thus obtained there are, in proportion to the intensity of the light, one, two, or more bright lines occupying precisely the same relative position. Moreover, very thin sheets of white light proceeding from various incandescent substances are passed through incandescent hydrogen, and the emergent light is then separated into its constituent elements by a prism. In the spectra thus obtained it is found that there are invariably (with the above-named exception) dark (or, under certain circumstances, bright[27]) lines occupying exactly the same positions in the spectrum as the lines above mentioned. Hence it is inferred, by the Method of Agreement, that a sheet of light, whether it proceed directly from incandescent hydrogen itself, or be transmitted through it from some other incandescent substance, will (allowing for the above exception) invariably produce these lines. But, if we try the same experiments with any other element than incandescent hydrogen, although we may obtain bright or dark lines, we never find these lines occupying the same positions in the spectrum as the lines in question.

Here, then, we have the negative instances of the Double Method; and it is inferred (subject, of course, to the assumption that our knowledge of the negative instances is sufficiently complete) that the presence in the spectrum of these lines is invariably due either to a sheet of light proceeding directly from incandescent hydrogen, or to a sheet transmitted through it from some other incandescent substance; that is to say, that one or other of these is the cause, and the *only* cause of the presence

---

[27] The darkness of the lines is due to the property possessed by incandescent media of absorbing sheets of light of the same refrangibility as those emitted by them. When the absorption exerted upon the transmitted light is more than compensated by the luminosity of the hydrogen light, these lines, instead of being dark, appear bright, as is also the case when the sheet of light proceeds directly from incandescent hydrogen itself. There is still a third case. When the hydrogen emits as much light as it absorbs, there will be no line, dark or bright.

in the spectrum of these particular lines. When these lines are bright, it is doubtful whether the rays proceed directly from incandescent hydrogen or have been transmitted through it, but, when they are dark, the sheets must have been transmitted. Wherever, therefore, dark lines occupying these positions occur in the spectrum we may infer (deductively) the passage of the sheet of light through a medium composed wholly or partially of incandescent hydrogen. But we detect such lines in the spectrum of the sun and several of the stars, and hence (unless we suppose it possible or not improbable that there is in the sun or other stars some element agreeing in this respect with hydrogen, but differing in others) we may conclude that the sun and these other stars are surrounded with an atmosphere of incandescent hydrogen [28].

The following examples are selected from a subject of a widely different character, the History of Language. M. Auguste Brachet, in his *Historical Grammar of the*

---

[28] It must be understood that, in this example, I have not stated the historical steps by which the discovery was arrived at, but simply attempted to give a logical analysis of the arguments by which it would now be established. It was the exact coincidence of the bright lines in the hydrogen spectrum with the dark lines in the solar spectrum, which first led to the belief that hydrogen enters into the constitution of the solar atmosphere. It is now, however, rendered possible, through an ingenious contrivance, to separate, as it were, the solar atmosphere from the glowing body within it, and thus to obtain these lines bright instead of dark. The student will find a brief account of these discoveries in Professor Stokes' Address to the British Association in 1869.

French Tongue [29], lays down the position that there are three sure tests by which we can discriminate between popular words derived from the Peasant Latin (lingua Latina rustica) by a regular process, and Latin words of learned origin imported into Modern French by scholars. These tests are (1) the continuance of the tonic accent; (2) the suppression of the short vowel; (3) the loss of the middle or medial consonant. It will be seen that it is by the employment of the Double Method of Agreement that M. Brachet arrives at these conclusions.

'Look at such words (i. e. words of popular origin) carefully, and you will see that the syllable accented in Latin continues to be so in French; or, in other words, that the accent remains where it was in Latin. This continuance of the accent is a general and absolute law: all words belonging to popular and real French respect the Latin accent: all such words as *portíque* from pórticus, or *viatíque* from viáticum, which break this law, will be found to be of learned origin, introduced into the language at a later time by men who were ignorant of the laws which nature had imposed on the transformation from Latin to French. We may lay it down as an infallible law, that *The Latin accent continues in French in all words of popular origin; all words which violate this law are of learned origin:* thus—

LATIN.	POPULAR WORDS.	LEARNED WORDS.
Alúmine	*alún*	*alumíne*
Ángelus	*ánge*	*angelús*
Blásphemum	*blâme*	*blasphéme*
Cáncer	*cháncre*	*cancér*
Cómputum	*cómpte*	*compút*
Débitum	*détte*	*débít*
Décima	*díme*	*décíme*, &c.

²⁹ Dr. Kitchin's Translation, p. 32; 7th ed. pp. 44–48.

‘ We have seen that the tonic accent is a sure touchstone by which
to distinguish popular from learned words. There is another means,
as certain, by which to recognise the age and origin of words—the
loss of the short vowel. Every Latin word, as we have said, is
made up of one accented vowel, and others not accented—one *tonic*
and others *atonic.* The tonic always remains; but of the atonics *the
short vowel, which immediately precedes the tonic vowel, always disappears in French:* as in—

Bon(ĭ)tátem	*bonté*
San(ĭ)tátem	*santé*
Pos(ĭ)túra	*posture*
Clar(ĭ)tátem	*clarté*
Sep(tĭ)mána	*semaine* (O. Fr. *sepmaine*)
Com(ĭ)tátus	*comté*
Pop(ŭ)látus	*peuplé,* &c.

‘ Words such as *circuler,* circuláre, which break this law and
keep the short vowel, are always of learned origin; all words of
popular origin lose it, as *cercler.* This will be seen from the following examples :—

LATIN.	POPULAR WORDS.	LEARNED WORDS.
Ang(ŭ)latus	*anglé*	*angulé*
Blasph(ĕ)máre	*blâmer* (O. Fr. *blasmer*)	*blasphémer*
Cap(ĭ)tále	*cheptel*	*capital*
Car(ĭ)tátem	*cherté*	*charité*
Circ(ŭ)láre	*cercler*	*circuler,* &c.

‘ The third characteristic, serving to distinguish popular from
learned words, is the loss of the medial consonant, i.e. of the consonant which stands between two vowels, like the t in matúrus.
We will at once give the law of this change :—*All French words
which drop the medial consonant are popular in origin, while words
of learned origin retain it.* Thus the Latin **vocalis** becomes, in
popular French, *voyelle,* in learned French, *vocale.* There are innumerable examples of this : as—

LATIN.	POPULAR WORDS.	LEARNED WORDS.
Au(g)ústus	*août*	*auguste*
Advo(c)átus	*avoué*	*avocat*

LATIN.	POPULAR WORDS.	LEARNED WORDS.
Anti(ph)óna	*antienne*	*antiphone*
Cre(d)éntia	*créance*	*crédence*
Communi(c)áre	*communier*	*communiquer*, &c.'

The requisitions of the Double Method of Agreement may be far from being rigorously fulfilled, and still two phenomena may be so frequently present together and so frequently absent together, that we may be justified in suspecting some causal connexion between them. Unless they were invariably absent together, as well as invariably present, and unless they were the *only* circumstances which were invariably present and absent together, we should not be justified in regarding one as the cause, and the *only* cause, of the other; but the mere detection of the fact that they are frequently present and absent together may justify us in believing that there is between them some causal connexion. The precise character of this causal connexion may hereafter be determined by one of the other inductive methods, or by bringing the case under a previous deduction. The following instances will serve as illustrations of what has been here said.

Sir John Herschel conceives that he has detected a connexion between a full moon and a calm night: 'The only effect distinctly connected with its [the moon's] position with regard to the sun, which can be reckoned upon with any degree of certainty, is its tendency to clear the sky of cloud, and to produce not only a serene but a *calm* night; when so near the full as to appear round to the eye—a tendency of which we have assured ourselves by long-continued and registered observation.' The precise nature of the causal connexion can here be determined : ' The effect in question, so far as the clearance of the sky is concerned, is traceable to a distinct physical cause, the warmth radiated from its [the full moon's] highly-heated surface; though why the effect should not continue for several nights after the full, remains problematic[30].'

In this example, there is not, of course, an *invariable* connexion between the clear night and the full moon ; for, in the determination of the weather, there are so many and so various causes at work that they must necessarily modify or counteract each other. The moon might exercise considerable influence, might, as Sir John Herschel says, have a *tendency* to produce a calm night and still be overpowered by other influences. It is suffi-. cient, in order to lead us to suspect some causal connexion between the two phenomena, that we should find a calm night proportionably oftener, and oftener in a considerable proportion, when there is a full moon than when there is not. Thus, suppose that, after a long series of observations of nights when there is a full moon, we find the proportion of calm nights to nights which cannot be called calm to be as 5 to 2 (I am, of course, taking an imaginary case), and the proportion on ordinary nights to be as 3 to 2, there can be little doubt

[30] Herschel's *Familiar Lectures on Scientific Subjects,* pp. 146. 147.

that the full moon is, in some way or other, connected with the larger proportion of calm nights.

The employment of the Double Method of Agreement may lead to the detection of facts of causation in many instances of a similar kind. Thus, suppose that, in a particular part of the country, a particular wind is found to be proportionably oftener attended with rain than other winds, we begin to suspect that there is some causal connexion between rain and this wind, so that, when the wind blows, we may expect rain, at least with more confidence than when other winds blow; and, if the proportion in which rain accompanies this wind be much greater than that in which it accompanies other winds, our expectation is proportionably strengthened, and we have no hesitation in speaking of the quarter from which the wind blows as ' the rainy quarter.' In this case, the cause is, of course, to be sought in the character of the tract over which the wind blows. Similarly, if, after sufficiently long observation, we find the death-rate in some particular place decidedly larger than in the surrounding neighbourhood, we have no hesitation in ascribing the fact to some peculiarity either in the place or the population, and we at once begin to consider whether there is anything exceptional in the soil, the climate, the habits or occupations of the people, and the like, which, either alone or in conjunction with other circumstances, would account for the phenomenon.

In all cases of this kind, we are, as it were, set on the track of a cause by discovering that some phenomenon is present in a proportionably greater number of instances when some other phenomenon is present than when it is absent. The cause itself may hereafter be detected either by one of the other inductive methods, or by bringing the case under a previous deduction. Thus, we know that the surface of that part of the moon which we call 'full' is highly heated, and that it is the tendency of warmth radiated from a highly-heated surface to clear the atmosphere. Hence the series of observed phenomena is, in this case, accounted for by being brought deductively under previous inductions.

METHOD OF RESIDUES.

CANON.

Subtract from any phenomenon such part as is known to be the effect of certain antecedents, and the residue of the phenomenon is the effect of the remaining antecedents.

If the antecedents are A, B, C, D, and the complex phenomenon can be resolved into the consequents a, β, γ, δ, ϵ, of which γ, δ, ϵ are ascertained by previous inductions or deductions to be due to C, D, then the remaining consequents a, β must be referred to the remaining antecedents A, B. Given that the total result is due to a certain number of antecedents, and that part of the result is due to a portion of those antecedents; the residue of the result must necessarily be due to the remaining antecedents. This rule appears so obvious as to be hardly worth stating; it has, however, as will be seen from the examples given below, been mainly instrumental in leading to many of the most important discoveries of modern times. 'It is by this process, in fact,' says Sir John Herschel[31], 'that science, in its present advanced state, is chiefly promoted. Most of the phenomena which nature presents are very complicated; and when the effects of all known causes are estimated with exactness, and subducted, the residual facts are constantly appearing in the form of phenomena altogether new, and leading to the most important conclusions.'

There is one difficulty connected with this Method. Subtraction being a deductive process, why is the Method of Residues included among the inductive methods? The Method, it must be confessed, is rather of a deductive than an inductive character, but as, in assigning the residual consequent to the residual antecedent, we assume the Law of Universal Causation, and as, moreover, the method is generally applied to the result of previous inductions and generally suggests subsequent inductions, it may vindicate its claim to discussion in this place. It is by induction that we usually ascertain that γ, δ, ϵ are due to C, D; by the Method of Residues we determine that the remaining consequents a, β must be due to the remaining antecedents A, B; we then generally proceed to decide by one of the other inductive methods which of the remaining consequents is due to which of the remaining antecedents.

The following are instances of the employment of the

[31] *Discourse on the Study of Natural Philosophy*, § 158.

Method of Residues, and it will be noticed that the science of astronomy is peculiarly rich in such examples [32] :—

'The planet Jupiter is attended by four satellites which revolve round it in orbits very nearly circular, and whose dimensions, forms, and situations with respect to that of the planet itself are now perfectly well known. The periodical times of their respective revolutions are also ascertained with extreme precision, and all the particulars of their motions have been investigated with extraordinary care and perseverance. The three interior of them are so near the planet, and the planes of their orbits so little inclined to that in which it revolves round the sun, that they pass through its shadow, and therefore undergo eclipse, at every revolution. These eclipses have been assiduously observed ever since the discovery of the satellites, and their times of occurrence registered. As they afford a means of determining the longitudes of places, the *prediction* beforehand of the exact times of their occurrence becomes an object of great importance: and it is evident enough that, all the particulars of their motions being known (as well as of that of the planet itself, and therefore of the size and situation of its shadow), there would be no difficulty in making such prediction (starting from the time of some one observed eclipse of each as an epoch); *provided always* each eclipse were *seen at the identical moment when it actually happened.* Moreover, on that supposition, the times *recorded* of all the subsequent

---

[32] In the first and second editions the acceleration (or diminution of the periodic times) of Encke's comet (see Herschel's *Discourse on the Study of Natural Philosophy*, § 159) was given as an example of residual phenomena. The cause of this phenomenon is, however, so doubtful, that I have thought it best to omit the instance in the later editions.

eclipses ought to agree with the times so *predicted*. This, however, proved not to be the case. The observed times were sometimes earlier, sometimes later than the predicted; not, however, capriciously, but according to a regular law of increase and decrease in the amount of discordance, the difference either way increasing to a maximum,—then diminishing, vanishing, and passing over to a maximum the other way, and the total amount of fluctuation to and fro being about $16^m\ 27^s$. Soon after this discrepancy between the predicted and observed times of eclipse was noticed, it was suggested that such a disagreement would necessarily arise if the transmission of light were not instantaneous. This suggestion was converted into a certainty by Roemer, a Danish astronomer, who ascertained that they always happened earlier than their calculated time when the earth in the course of its annual revolution approached nearest to Jupiter, and later when receding farthest: so that in effect the extreme difference of the errors or total extent of fluctuation—the $16^m\ 27^s$ in question—is no other than the time taken by light to travel over the diameter of the earth's orbit, that being the extreme difference of the distances of the two planets at different points of their respective revolutions. At present, in our almanacs a due allowance of time for the transmission of light at this rate, assuming a uniform velocity, is made in the calculation of these eclipses; and the discrepancy in question between the observed and predicted times has ceased to exist[33].'

The circumstances which led to the discovery of the planet Neptune furnish, perhaps, the most striking instance of the employment of the Method of Residues. From the year 1804 it had been noticed that the orbit of the planet Uranus was subject to an amount of per-

[33] Herschel's *Familiar Lectures on Scientific Subjects*, p. 226, &c.

turbation which could not be accounted for from the influence of the known planets.

'Of the various hypotheses formed to account for it (the perturbation), during the progress of its development, none seemed to have any degree of rational probability but that of the existence of an exterior, and hitherto undiscovered, planet, disturbing, according to the received laws of planetary disturbance, the motion of Uranus by its attraction, or rather superposing its disturbance on those produced by Jupiter and Saturn, the only two of the old planets which exercise any sensible disturbing action on that planet. Accordingly, this was the explanation which naturally, and almost of necessity, suggested itself to those conversant with the planetary perturbations who considered the subject with any degree of attention. The idea, however, of setting out from the observed anomalous deviations, and employing them as data to ascertain the distance and situation of the unknown body, or, in other words, to resolve the inverse problem of perturbations, "*given the disturbances to find the orbit and place in that orbit of the disturbing planet*," appears to have occurred only to two mathematicians, Mr. Adams in England and M. Leverrier in France, with sufficient distinctness and hopefulness of success to induce them to attempt its solution. Both succeeded, and their solutions, arrived at with perfect independence, and by each in entire ignorance of the other's attempt, were found to agree in a surprising manner when the nature and difficulty of the problem is considered; the calculations of M. Leverrier assigning for the heliocentric longitude of the disturbing planet for the 23rd Sept. 1846, 326° 0', and those of Mr. Adams (brought to the same date) 329° 19', differing only 3° 19'; the plane of its orbit deviating very slightly, if at all, from that of the ecliptic.

' On the day above mentioned—a day for ever memorable in the annals of astronomy—Dr. Galle, one of the astronomers of the Royal Observatory at Berlin, received a letter from M. Leverrier, announcing to him the result he had arrived at, and requesting him to look for the disturbing planet in or near the place assigned by his calculation. He did so, and *on that very night actually found it.* A star of the eighth magnitude was seen by him and by M. Encke in a situation where no star was marked as existing in Dr. Bremiker's chart, then recently published by the Berlin Academy. The next night it was found to have moved from its place, and was therefore assuredly a planet. Subsequent observations and calculations have fully demonstrated this planet, to which the name of Neptune has been assigned, to be really that body to whose disturbing attraction, according to the Newtonian law of gravity, the observed anomalies in the motion of Uranus were owing[34].'

Besides furnishing an instance of the method of Residues, the above example is also a happy illustration of the combination of deduction with observation which has been so eminently fertile in astronomical research.

'Almost all the greatest discoveries in astronomy have resulted from the consideration of what we have elsewhere termed RESIDUAL PHENOMENA, of a quantitative or numerical kind, that is to say, of such portions of the numerical or quantitative results of observation as remain outstanding and unaccounted for after subducting and allowing for all that would result from the strict application of known principles. It was thus that the grand discovery of the precession of the equinoxes resulted, as a residual phenomenon, from the imperfect explanation of the return of the seasons by the return of the sun to the same apparent place among the fixed stars. Thus, also, aberration and nutation resulted as residual phenomena from that portion of the changes of the apparent

[34] Herschel's *Outlines of Astronomy*, Fourth Edition, §§ 767, 768.

places of the fixed stars which are left unaccounted for by precession. And thus again the *apparent* proper motions of the stars are the observed *residues* of their apparent movements outstanding and unaccounted for by strict calculation of the effects of precession, nutation, and aberration. The nearest approach which human theories can make to perfection is to diminish this residue, this *caput mortuum* of observation, as it may be considered, as much as practicable, and, if possible, to reduce it to nothing, either by showing that something has been neglected in our estimation of known causes, or by reasoning upon it as a new fact, and on the principle of the inductive philosophy ascending from the effect to its cause or causes[35].'

' Many of the new elements of chemistry have been detected in the investigation of *residual phenomena.* Thus, Arfwedson discovered lithia by perceiving an *excess of weight* in the sulphate produced from a small portion of what he considered as magnesia present in a mineral he had analysed. It is on this principle, too, that the *small concentrated residues of great operations* in the arts are almost sure to be the lurking places of new chemical ingredients : witness iodine, brome, selenium, and the new metals accompanying platina in the experiments of Wollaston and Tennant. It was a happy thought of Glauber to examine what everybody else threw away[36].'

' The unforeseen effects of changes in legislation, or of improvements in the useful arts, may often be discerned by the Method of Residues. In comparing statistical accounts, for example, or other registers of facts, for a series of years, we perceive at a certain period an altered state of circumstances, which is unexplained by the ordinary course of events, but which must have some cause. For this *residuary*

[35] Herschel's *Outlines of Astronomy*, § 856.
[36] Herschel's *Discourse on the Study of Natural Philosophy*, § 161.

phenomenon, we seek an explanation until it is furnished by the incidental operation of some collateral cause. For example, on comparing the accounts of live cattle and sheep annually sold in Smithfield market for some years past, it appears that there is a large increase in cattle, while the sheep are nearly stationary. The consumption of meat in London may be presumed to have increased, at least in proportion to the increase of its population; and there is no reason for supposing that the consumption of beef has increased faster than that of mutton. There is, therefore, a residuary phenomenon—viz. the stationary numbers of the sheep sold in Smithfield—for which we have to find a cause. This cause is the increased transport of dead meat to the metropolis, owing to steam navigation and railways, and the greater convenience of sending mutton than beef in a slaughtered state.

'Again: on comparing the consumption of wine with that of spirits and beer in England during the last sixty years[37], we find that the former has remained stationary, while the latter has undergone a great increase. The general causes, such as increase of population and wealth, which have increased the consumption of spirits and beer, have not increased the consumption of wine. For this residuary phenomenon, a special cause must be sought; and it may be found principally in the alteration of habits among the upper classes with respect to drinking[38].'

[37] This passage was written in 1852. Since that time, owing to the reduction of the duties, the greater familiarity of Englishmen with foreign countries and habits, and, perhaps, the taste for a more refined style of living, the consumption of wine has enormously increased (First Edition, 1870). On the other hand, owing to the spread of Temperance Societies, and of more temperate habits in all classes of the population, the consumption of all alcoholic drinks has now, for some years, been steadily diminishing.

[38] Sir George Cornewall Lewis on the *Methods of Observation and Reasoning in Politics*, vol. i. p. 356.

I shall conclude my instances with what Sir John Herschel truly calls 'a very elegant example,' the difference between the observed and calculated velocities of sound. I quote from Professor Tyndall's *Lectures on Sound:*—

'I now come to one of the most delicate points in the whole theory of sound. The velocity through air has been determined by direct experiment; but knowing the elasticity and density of the air, it is possible without any experiment at all to calculate the velocity with which a sound-wave is transmitted through it. Sir Isaac Newton made this calculation, and found the velocity at the freezing temperature to be 916 feet a second. This is about one-sixth less than actual observation had proved the velocity to be, and the most curious suppositions were made to account for the discrepancy. Newton himself threw out the conjecture that it was only in passing from particle to particle of the air that sound required *time* for its transmission; that it moved instantaneously *through the particles themselves.* He then supposed the line along which sound passes to be occupied by air-particles for one-sixth of its extent, and thus he sought to make good the missing velocity. The very art and ingenuity of this assumption were sufficient to ensure its rejection; other theories were therefore advanced, but the great French mathematician Laplace was the first to completely solve the enigma. I shall now endeavour to make you thoroughly acquainted with his solution.

'I hold in my hand a strong cylinder of glass, accurately bored, and quite smooth within. Into this cylinder, which is closed at the bottom, fits this air-tight piston. By pushing the piston down, I condense the air beneath it; and, when I do so, heat is developed. Attaching a scrap of amadou to the bottom of the piston, I can ignite it by the heat generated by compression. Dipping a bit of cotton wool into bisulphide of carbon, and attaching it to the piston, when the latter is forced down, a flash of light, due to the ignition of the bisulphide of carbon vapour, is observed within the tube. It is thus proved that, when air is compressed, heat is generated. By another experiment, I can show you that, when air is rarefied, cold is developed. This brass box contains a quantity of condensed air. I open the cock, and permit the air to discharge itself against a suitable thermometer; the sinking of the instrument declares the chilling of the air.

'All that you have heard regarding the transmission of a sonorous pulse through air, is, I trust, still fresh in your minds. As the pulse advances, it squeezes the particles of air together, and two results follow from this compression of the air. Firstly, its elasticity is augmented through the mere augmentation of its density. Secondly, its elasticity is augmented by the heat developed by compression. It was the change of elasticity which resulted from a change of density that Newton took into account, and he entirely overlooked the augmentation of elasticity due to the second cause above mentioned. Over and above, then, the elasticity involved in Newton's calculation, we have an additional elasticity due to changes of temperature produced by the sound itself. When both are taken into account, the calculated and the observed velocity agree perfectly[39].'

. It is not necessary, for our purposes, to pursue the quotation, but the student, who wishes to see an example of the extreme delicacy and caution with which it is requisite to conduct physical researches, may with great advantage read the remainder of the chapter.

[39] *Lectures on Sound*, ch. i.

METHOD OF CONCOMITANT VARIATIONS.

CANON.

Whatever phenomenon varies in any manner whenever another phenomenon varies in some particular manner, is either a cause or an effect of that phenomenon, or is connected with it through some fact of causation [40].

This Method is really a peculiar application, or series of applications, of the Method of Difference. It is employed in those cases where a phenomenon cannot be made to disappear altogether, but where we have the power of augmenting or diminishing its quantity, or at least where Nature presents it in greater or smaller amounts. Thus, suppose we drop a quantity of quicksilver into a glass tube, we shall find that every sensible augmentation of the temperature of the surrounding atmosphere is accompanied by a sensible augmentation of the volume of the quicksilver in the tube, and, *vice versâ*, that every sensible diminution of the temperature is accompanied by a sensible diminution of the volume of the quicksilver. Now each particular experiment is an application of the Method of Difference, and, providing we have ascertained that the conditions of that Method have been rigorously satisfied, partakes of its cogency. That certain definite augmentations of temperature will increase the volume of quicksilver by, say, one-twentieth, one-thirtieth, or one-fiftieth part, is an

[40] On p. 186 will be found a rider to this Canon.

absolutely certain inference, supposing due care to have been taken in the performance of the experiments, and is simply a result of the Method of Difference. But, inasmuch as there are limits above and below which we cannot try the experiment, or intermediate points of temperature at which we do not find it convenient to do so, the question arises whether we are justified, in virtue of the experiments already tried, in asserting that the volume of the quicksilver will invariably expand or contract in proportion to the increasing or diminishing temperature of the surrounding media. We are justified in making such an assertion, and for this reason. The cause which occasions the quicksilver to expand or contract at two definite points must, if it continue to act, and if it be counteracted by no other cause, operate at all intermediate points; and, similarly, the cause which occasions it to expand or contract up to a certain point must, on the same suppositions, go on producing a like effect. This fact is implied in the very notion of Causation. We arrive, then, at the conclusion that the volume of the quicksilver is invariably dependent on the temperature of the surrounding medium; in other words, that augmentation of temperature is *the cause* of its expansion [41].

It may be asked, Why not employ the Method of

[41] The student acquainted with the phraseology of Mathematics will understand my meaning, when I say that the Method of Concomitant Variations is really an integration of a (supposed) infinite number of applications of the Method of Difference.

Difference once and for all? Because, *ex hypothesi*, the phenomenon is one which is only capable of augmentation or diminution, and cannot be made to vanish. We may reduce to a minimum the resistance to motion, but we cannot remove the resistance altogether. We may more and more diminish the heat of a body, but we cannot wholly deprive the body of its heat. Hence we can apply the Method of Difference to the several augmentations and diminutions of the phenomenon, but we cannot apply it to the phenomenon as a whole.

In the example given above, we know that augmentation of temperature and augmentation of volume are related as cause and effect, because, in the experiments, we can assure ourselves that they are the only two circumstances which vary in common; if we were not certain of this fact, there might be some third circumstance which was the cause of both. Moreover, we know that augmentation of temperature is the cause and augmentation of volume the effect, because, in this case, the former is the antecedent and the latter the consequent. There is another class of cases where, though we are not able to determine which of two circumstances is cause and which is effect, we may regard the *relation* as being one of cause and effect, inasmuch as we feel confident that there is present no third circumstance varying proportionately with the other two. But, if we cannot be confident even of this fact, the two circumstances may, for aught we know, both be effects of the same cause; as, for instance, the loudness of a clap of thunder varies with the intensity of a flash of lightning, though neither is the cause of the other, both alike being effects of the electrical condition of the atmosphere.   Hence will be seen the importance of the concluding words of the Canon, 'or is connected with it through some fact of causation.'   The first and second cases differ from the third in this, that in both of them we suppose a rigorous fulfilment of the requisitions of the Method of Difference as applied to those individual observations or experiments on which the Method of Concomitant Variations is founded, and which it, as it were, sums up.   In the last case, however, we suppose that there is some uncertainty as to whether the requisitions of the Method of Difference have been rigorously fulfilled or not.   It will thus be seen that the statement of the Canon, as given above, is adapted to the weakest case.   I may add to it the following rider :—

If we can assure ourselves that there is no third phenomenon varying concurrently with these two, we may affirm that the one phenomenon is either the cause or the effect of the other.

The Method of Concomitant Variations may be used for two purposes, either to establish a causal connexion, or to determine the law according to which two phenomena vary.   Thus, it may either establish the fact that any increase of temperature causes quicksilver to expand, or it may determine the exact rate according to which this expansion takes place, a determination which is, in fact, effected by the ordinary thermometer.   In the latter case, the application of the Method is not confined to those permanent natural agents referred to above, the influence of which we cannot altogether remove; it may come in as supplementary to the Method of Difference. Thus it is by the Method of Difference that we discover that certain kinds of impurity in the atmosphere produce certain kinds of disease, but, if we could ascertain the relation subsisting between the amount of impurity in the atmosphere and the amount of disease, it would be by an application of the Method of Concomitant Variations.

In the latter class of enquiries, the attempt to determine the numerical relations according to which two phenomena vary, the utmost caution is required as soon as our inference outsteps the limits of our observations. In the first place, there is always the possibility of the intervention of some counteracting cause. In the case of water, we found that, at 39°, instead of continuing to contract as it becomes colder, it ceases at that point to do so, and thenceforward begins to expand. 'No counteracting cause intervening' is, however, a qualification with which we must understand all our inductions, by whatever method they may have been arrived at. But there is an element of uncertainty which is peculiar to the Method of Concomitant Variations as applied to determine the *law* or *rate* of variation between two phenomena, especially when the range of our observations is confined within comparatively narrow limits. 'Any one,' says Mr. Mill [42], 'who has the slightest acquaint-

[42] Mill's *Logic*, Bk. III. ch. viii. § 7.

ance with mathematics, is aware that very different laws of variation may produce numerical results which differ but slightly from one another within narrow limits; and it is often only when the absolute amounts of variation are considerable, that the difference between the results given by one law and by another becomes appreciable. When, therefore, such variations in the quantity of the antecedents as we have the means of observing are small in comparison with the total quantities, there is much danger lest we should mistake the numerical law, and be led to miscalculate the variations which would take place beyond the limits; a miscalculation which would vitiate any conclusion respecting the dependence of the effect upon the cause, that could be founded on those variations. Examples are not wanting of such mistakes. "The formulæ," says Sir John Herschel, "which have been empirically deduced for the elasticity of steam (till very recently), and those for the resistance of fluids, and other similar subjects," when relied on beyond the limits of the observations from which they were deduced, "have almost invariably failed to support the theoretical structures which have been erected on them."' This, however, it must be noticed, is an uncertainty which does not vitiate the method, but simply renders necessary the exercise of the utmost caution in its application.

Perhaps no simpler instance of the Method of Concomitant Variations can be given than the experimental proof of the First Law of Motion, which Law may be stated thus: that all bodies in motion continue to move in a straight line with uniform velocity until acted upon by some new force.

'This assertion,' I am quoting from Mr. Mill[43], 'is in open opposition to first appearances ; all terrestrial objects, when in motion, gradually abate their velocity and at last stop ; which accordingly the ancients, with their *inductio per enumerationem simplicem*, imagined to be the law. Every moving body, however, encounters various obstacles, as friction, the resistance of the atmosphere, &c., which we know by daily experience to be causes capable of destroying motion. It was suggested that the whole of the retardation might be owing to these causes. How was this enquired into ? If the obstacles could have been entirely removed, the case would have been amenable to the Method of Difference. They could not be removed, they could only be diminished, and the case, therefore, admitted only of the Method of Concomitant Variations. This accordingly being employed, it was found that every diminution of the obstacles diminished the retardation of the motion : and, inasmuch as in this case (unlike the case of heat) the total quantities both of the antecedent and of the consequent were known, it was practicable to estimate, with an approach to accuracy, both the amount of the retardation and the amount of the retarding causes or resistances, and to judge how near they both were to being exhausted ; and it appeared that the effect dwindled as rapidly [as the cause], and at each step was as far on the road towards annihilation as the cause was. The simple oscillation of a weight suspended from a fixed point, and moved a little out of the perpendicular, which in ordinary circumstances lasts but a few minutes, was prolonged in Borda's experiments to more than thirty hours, by diminishing as much as possible the friction at the point of suspension, and by making the body oscillate in a space exhausted as

[43] Mill's *Logic*, Bk. III. ch. viii. § 7.

nearly as possible of its air. There could therefore be no hesitation in assigning the whole of the retardation of motion to the influence of the obstacles : and since, after subducting this retardation from the total phenomenon, the remainder was an uniform velocity, the result was the proposition known as the first law of motion.'

I have already employed as an illustration the fact that a change in the temperature of a body is always accompanied by a change in its volume. The following extract places the same fact in a new light, and shows that the nature of substance (whether solid, liquid, or aëriform) depends on, and, at considerable intervals, varies with, the temperature to which it is exposed.

'The most striking and important of the effects of heat consist, however, in the liquefaction of solid substances, and the conversion of the liquids so produced into vapour. There is no solid substance known which, by a sufficiently intense heat, may not be melted, and finally dissipated in vapour ; and this analogy is so extensive and cogent, that we cannot but suppose that all those bodies, which are liquid under ordinary circumstances, owe their liquidity to heat, and would freeze or become solid if their heat could be sufficiently reduced. In many we see this to be the case in ordinary winters ; for some, severe frosts are requisite ; others freeze only with the most intense artificial colds ; and some have hitherto resisted all our endeavours ; yet the number of these last is few, and they will probably cease to be exceptions as our means of producing cold become enlarged.

'A similar analogy leads us to conclude that all aëriform fluids are merely liquids kept in the state of vapour by heat. Many of them have been actually condensed into the liquid state by cold accompanied with violent pressure ; and, as our means of applying these causes of condensation have improved, more and more refractory ones have successively yielded. Hence we are fairly entitled to extend our conclusion to those which we have not yet been able to succeed with ; and thus we are led to regard it as a general fact, that the liquid and aëriform or vaporous states are entirely dependent on *heat;* that, were it not for this cause, there would be nothing but solids in nature ; and that, on the other hand, nothing but a sufficient intensity of heat is requisite to destroy the cohesion of every substance, and reduce all bodies, first to liquids, and then into vapour[44].'

An interesting application of the Method of Concomitant Variations is found in the arguments by which it is established that refrigeration *at night* (when the sun's rays are withdrawn) is, *cæteris paribus*, proportional to the *dryness* of the atmosphere. Thus, in the British Isles, where the atmosphere almost always contains a large amount of aqueous vapour, the difference between

[44] Herschel's *Discourse on the Study of Natural Philosophy*, §§ 357, 358.

Sir John Herschel's conjecture has been verified with regard to aëriform fluids, but not as yet with regard to liquids. The experiments of Cailletet and Pictet (an account of which may be found in the *Academy* of Jan. 12th, 1878, and in *Nature* of Jan. 3rd and 17th of the same year) have conclusively shown that even oxygen, hydrogen, and nitrogen admit of liquefaction, and, therefore, probably of solidification. A brief account of these experiments is given in Ganot's *Physics*, Translation, 12th ed. § 382. Thus, the old distinction between permanent and non-permanent gases has been entirely effaced. By a legitimate analogy, it may be inferred, with a very high degree of probability, that all liquids admit of solidification. But some liquids, such as alcohol, ether, and bisulphide of carbon, have hitherto resisted all attempts to solidify them, even at the lowest known temperature. See Ganot, § 343.

the temperature at day and night is comparatively slight, whereas, in countries far inland, where the atmosphere is extremely dry, the variations of temperature are comparatively large. This phenomenon is due to the fact that masses of aqueous vapour, though they intercept, also radiate heat. Hence, while during the day they protect us from the excessive heat of the sun, they intercept the heat which is radiated from the earth's surface during the night, and, at the same time, return to it some portion of the heat they have absorbed during the day.

'A freedom of escape,' says Professor Tyndall[45], 'would occur at the earth's surface generally, were the aqueous vapour removed from the air above it, for the great body of the atmosphere is a practical vacuum, as regards the transmission of radiant heat. The withdrawal of the sun from any region over which the atmosphere is dry, must be followed by quick refrigeration. The moon would be rendered entirely uninhabitable by beings like ourselves through the operation of this single cause ; with a radiation, uninterrupted by aqueous vapour, the difference between her monthly maxima and minima must be enormous. The winters of Thibet are almost unendurable, from the same cause. Witness how the isothermal lines dip from the north into Asia, in winter, as a proof of the low temperature of this region. Humboldt has dwelt upon the " frigorific power" of the central portions of this continent, and controverted the idea that it was to be explained by reference to the elevation ; there being vast expanses of country, not much above the sea-level, with an exceedingly low temperature. But, not knowing the influence which we are now studying, Humboldt, I imagine, omitted the most potent cause of the cold. The

[45] Tyndall's *Heat a Mode of Motion,* § 492.

refrigeration at night is extreme when the air is dry. The removal, for a single summer night, of the aqueous vapour from the atmosphere which covers England, would be attended by the destruction of every plant which a freezing temperature could kill. In Sahara, where "the soil is fire and the wind is flame," the cold at night is often painful to bear. Ice has been formed in this region at night. In Australia, also, the *diurnal range* of temperature is very great, amounting, commonly, to between 40 and 50 degrees. In short, it may be safely predicted that, wherever the air is *dry*, the daily thermometric range will be great. This, how-ever, is quite different from saying that, where the air is *clear*, the thermometric range will be great. Great clearness to light is perfectly compatible with great opacity to heat ; the atmosphere may be charged with aqueous vapour while a deep blue sky is overhead, and on such occasions the terrestrial radiation would, notwithstanding the "clearness," be intercepted.'

The science of Geology abounds in instances of the employment of the Method of Concomitant Variations. In fact, as the agents with which it is concerned, land and water, subsidence and elevation, deposition and denudation, are constantly present and acting on the earth's surface, and as it is impossible to cause the influence of any one of them to vanish altogether, the geologist is compelled in his explanations and arguments to avail himself mainly of this method. The following extract from Lyell's *Principles of Geology* furnishes a good illustration, and will be peculiarly interesting to any one who has visited the place. It is designed as an explanation of the alternate subsidence and elevation of the famous temple of Jupiter Serapis, at Pozzuoli, on the Bay of Naples.

'We can scarcely avoid the conclusion, as Mr. Babbage has hinted, "that the action of heat is in some way or other the cause of the phenomena of the change of level of the temple. Its own hot spring, its immediate contiguity to the Solfatara, its nearness to the Monte Nuovo, the hot spring at the baths of Nero on the opposite side of the Bay of Baiæ, the boiling springs and ancient volcanos of Ischia on one side and Vesuvius on the other, are the most prominent of a multitude of facts which point to that conclusion." And when we reflect on the dates of the principal oscillations of level, and the volcanic history of the country before described, we seem to discover a connexion between each era of upheaval and a local development of volcanic heat, and again between each era of depression and the local quiescence or dormant condition of the subterranean igneous causes. Thus for example, before the Christian era, when so many vents were in frequent eruption in Ischia, and when Avernus and other points in the Phlegræan Fields were celebrated for their volcanic aspect and character, the ground on which the temple stood was several feet above water. Vesuvius was then regarded as a spent volcano; but when, after the Christian era, the fires of that mountain were rekindled, scarcely a single outburst was ever witnessed in Ischia, or around the Bay of Baiæ. Then the temple was sinking. Vesuvius, at a subsequent period, became nearly dormant for five centuries preceding the great outbreak of 1631, and in that interval the Solfatara was in eruption A.D. 1198, Ischia in 1302, and Monte Nuovo was formed in 1538. Then the foundations on which the temple stood were rising again. Lastly, Vesuvius once more became a most active vent, and has been so ever since, and during the same lapse of time the area of the temple, so far as we know anything of its history, has been subsiding.

'These phenomena would agree well with the hypothesis, that when the subterranean heat is on the increase, and when lava is forming without obtaining an easy vent, like that afforded by a great habitual chimney, such as Vesuvius, the incumbent surface is uplifted, but when the heated rocks below are cooling and contracting, and sheets of lava are slowly consolidating and diminishing in volume, then the incumbent land subsides[46].'

Laplace's *Nebular Hypothesis*, that stellar systems, like our solar system, are formed from the gradual condensation of nebular masses, is supported by an appeal to this method. 'We see,' conceives Laplace, 'among these nebulæ' (which are diffused along the Milky Way), 'instances of all degrees of condensation, from the most loosely diffused fluid, to that separation and solidification of parts by which suns and satellites and planets are formed : and thus we have before us instances of systems in all their stages; as in a forest we see trees in every period of growth [47].'

Physiology (so far as it is based on Anatomy, as distinct from direct experiment), for like reasons with Geology, mainly employs the Method of Concomitant Variations. It is very seldom, in this science, that we obtain a phenomenon present in one set of instances

[46] Lyell's *Principles of Geology*, tenth edition, ch. xxx.

[47] Whewell's *Novum Organon Renovatum*, Bk. III. ch. viii. sect. 2. § 9. This example is adduced by Dr. Whewell as an instance of what he calls the Method of Gradation, which, however, must not be confounded with Mill's Method of Concomitant Variations. The example, so far as it can be relied on, serves equally well as an instance of either method.

and entirely absent in another; but we frequently find a phenomenon which, within certain limits, presents itself in the most variable quantities. If, then, we find another phenomenon varying as it varies, we may argue with tolerable confidence that the two phenomena either stand to each other in the relation of cause and effect, or are, at least, common effects of some unknown cause. Thus, it appears to be established that, not only in different species, but in different individuals of the same species, there is some relation between the manifestations of intelligence and the amount of cerebral development, understanding the latter expression to include not only bulk of brain but also complexity and depth of convolutions.

'With some apparent exceptions,' says Dr. Carpenter[48], a classical authority on most physiological questions, 'which there would probably be no great difficulty in explaining if we were in possession of all the requisite data, there is a very close correspondence between the relative development of the Cerebrum in the several tribes of Vertebrata and the degree of Intelligence they respectively possess—using the latter term as a comprehensive expression for that series of mental actions which consists in the *intentional* adaptation of means to ends, based on definite *ideas* as to the nature of both.'

And again :—

'As we ascend the Mammalian series, we find the Cerebrum becoming more and more elongated posteriorly by the development of the middle lobes, and the intercerebral conmissure becomes more complete; but we must ascend as

[48] Carpenter's *Principles of Human Physiology*, sixth edition, 1864.

high as the Carnivora, before we find the least vestige of the posterior lobes; and the rudiment which these possess is so rapidly enlarged in the Quadrumana, that in some of that group the posterior lobes are as fully developed in reference to the Cerebrum as a whole, and as completely cover in the Cerebellum, as in the human subject. The attention which has yet been given to this department of enquiry, has not hitherto done more than confirm the statement already made, with regard to the general correspondence between the development of the Cerebrum and the manifestations of Intelligence; very decided evidence of which is furnished by the great enlargement of the Cerebrum, and the corresponding alteration in the form of the Cranium, which present themselves in those races of Dogs most distinguished for their educability, when compared with those whose condition approximates most closely to what was probably their original state of wildness.

'This general inference, drawn from Comparative Anatomy, is borne out by observation of the human species. When the Cerebrum is fully developed, it offers innumerable diversities of form and size among various individuals; and there are as many diversities of character. It may be doubted if two individuals were ever exactly alike in this respect. That a Cerebrum which is greatly under the average size is incapable of performing its proper functions, and that the possessor of it must necessarily be more or less idiotic, there can be no reasonable doubt. On the other hand, that a large, well-developed Cerebrum is found to exist in persons who have made themselves conspicuous in the world in virtue of their intellectual achievements, may be stated as a proposition of equal generality.'

Dr. Thurnam [49], taking the brain-weights of ten dis-

[49] *On the Weight of the Brain,* by John Thurnam, M.D. London, J. E. Adlard. 1866.

tinguished men, who died between the ages of fifty and seventy, calculates the average weight of their brains to have been 54·7 ounces. The average weight of the brains of ordinary men, dying between the same ages, is 47·1 ounces. These facts give in favour of 'cultivated and intellectual man' an excess of 7·6 ounces, or 15 per cent. Though, as a general rule, the connexion between intellectual and cerebral development appears to be substantiated, we must, however, be very cautious in drawing any inferences as to particular cases. Megalocephaly, or pathological enlargement of the brain, is a recognised disease, and is frequently attended with idiotcy. In this class of cases, no doubt, if our means of investigation were adequate, we should discover some peculiarity either in the chemical composition or in the anatomical structure of the brain which would enable us to explain the exceptions in conformity with the rule.

It is, perhaps, needless to add that we are not justified in drawing any further inference from these data, than that the brain is the *organ* of intelligence, and that there is some definite relation between the organ and its functions.

Another interesting application of the Method of Concomitant Variations may be found in one of the arguments by which the distinction between Formed and Germinal Material is established. Any piece of glandular tissue, if examined under a microscope, will be found to consist of two parts, one of which will take a tint from carmine, the other not. The portion which takes the tint is called Germinal, the portion which will not take it is called Formed Material. The former is living matter, capable of growth and germination; the latter is dead matter, capable of no change but decay. Now, if this distinction between the two kinds of matter be well founded, we may reasonably expect to find the germinal matter developed in much larger proportions in the younger than in the older specimens of animals and plants, and in what may be called the more active than in what may be called the more passive parts of animal and vegetable organisms. And this is the case. In the pith of rush, elder, &c. we find that, in the spring, there are many portions of the cells which will take the carmine tint; in summer, few; in autumn, none. In the crystalline lens of the eyes of young animals the portions which will take it preponderate, becoming proportionately fewer as we examine the eyes of older specimens. In the grey matter of the brain we find many parts which will take the carmine tint, in the white matter but few. In a grain of wheat, when formed, there is, in the perisperm, no portion which will take it, in the white matter but a small portion, while in the embryo it is often difficult to discover any part which does not take it. These instances might be multiplied to any extent [50].

In physiological and medical researches, we must be peculiarly careful to bear in mind that, though two pheno-

[50] The student will find this subject fully treated in Dr. Lionel Beale's *Lectures on the Simple Tissue of the Human Body*, and in other works of the same author.

mena may vary proportionately, it by no means follows that one is cause and the other effect. They may both be common effects of the same cause. Thus, though the prevalence of cholera is 'said to be constantly attended by the appearance of certain low forms of organic life, namely, fungi or phytozoa, it by no means follows that these fungi or phytozoa are the cause of cholera. Both phenomena alike may be effects of certain conditions of the atmosphere. Nothing but direct experiment can determine between these two theories.

The Method of Concomitant Variations is that which is most frequently employed in the Science of Language. It is found, for instance, that between two dissimilar words employed at different epochs to express the same idea may be interpolated a number of intermediate forms employed at intermediate epochs, which make the transition from the one word to the other gradual and natural. From this circumstance it is inferred that the word used at the later epoch is derived from that used at the earlier epoch, certain tendencies of speech being regarded as the cause of the divergence. 'Thus, at first sight,' says M. Brachet[51], 'it is hard to see that *âme* is derived from *anima*; but history, our guiding-line, shows us that in the thirteenth century the word was written *anme*, in the eleventh *aneme*, in the tenth *anime*, which leads us straight to the Latin *anima.*' In this case there can be no doubt of the truth of the conclusion.

[51] M. Brachet's *Historical Grammar of the French Tongue*, Dr. Kitchin's Translation, p. 42. Seventh Edition, p 53.

Similarly, the loss of declension in the transition from the Latin Language to the French is easily explained when we take into account the following considerations :—

'The tendency to simplify and reduce the number of cases was early felt in the popular Latin: the cases expressed shades of thought too delicate and subtle for the coarse mind of the Barbarian. And so, being unable to handle the learned and complicated machinery of the Latin declensions, he constructed a system of his own, simplifying its springs, and reducing the number of the effects at the price of frequently reproducing the same form. Thus the Roman distinguished by means of case-terminations the place where one is, from the place to which one is going : "veniunt ad domum," "sunt in domo." But the Barbarian, unable to grasp these finer shades, saw no use in this distinction, and said, in either case alike, "sum in domum," "venio ad domum."

'Thus, from the fifth century downwards, long before the first written records of the French language, popular Latin reduced the number of cases to two : (1) the nominative to mark the subject; and (2) that case which occurred most frequently in conversation, the accusative, to mark the object or relation. From that time onwards the Latin declension was reduced to this :—subject, *murus;* object, *murum.*

'The French language is the product of the slow development of popular Latin; and French grammar, which was originally nothing but a continuation of the Latin grammar, inherited, and in fact possessed from its infancy, a completely regular declension : subject, *murs, murus;* object, *mur, murum :* and people said " ce *murs* est haut;" "j'ai construit un *mur.*"

'This declension in two cases forms the exact difference between ancient and modern French. It disappeared in the fourteenth century, not without leaving many traces in the language, which look like so many insoluble exceptions, but find their explanation and historic justification in our knowledge of the Old French declension[52].'

Here the conclusion is that French Grammar is derived from Latin Grammar, certain peculiarities of the period intervening between the use of the Latin and modern French languages being regarded as the cause of the differences between them.

Again, nothing at first sight would appear more improbable than that the French word *suis* and the Greek word εἰμι are derived from the same root. But, when we compare the old French word *sui*, the Latin *sum*, the Old Latin *esum*, and the Old Greek form ἐσμι, the connexion of the two words and their ultimate derivation from a common root becomes a certainty. Here the divergence may be definitely accounted for by the various influences operating upon people (like the Latins and Greeks) occupying different tracts of country, exposed to different circumstances, having the organs of speech differently modified, and the like.

Amongst the above examples it will be noticed that some have been included, the conclusions of which are by no means absolutely certain. In these cases, the deficiency of proof is due not to any formal inconclusiveness in the Method of Concomitant Variations, or in that of Difference, on which it is based, but to the existence of a doubt as to whether the requirements of those

[52] M. Brachet's *Historical Grammar of the French Tongue*, Dr. Kitchin's Translation, p. 88. Seventh Edition, pp. 98–100.

methods have been stringently fulfilled. In any but the Experimental Sciences it is always extremely difficult to assure ourselves that we are acquainted with all the circumstances which may influence, or may be influenced by, any given phenomenon. Moreover, as is the case, for instance, with regard to the concomitance between cerebral development and the manifestation of intelligence, there may be many known points of difference between the observed cases besides those which are taken into account, and the value of the conclusion will depend on the extent to which we have ascertained that these other points of difference are not pertinent, or not equally pertinent with those which we have taken into account, to the circumstance or circumstances which we are investigating.

The application of the Method of Concomitant Variations to determine the numerical relations subsisting between two phenomena may be illustrated from the experiments by which the *measure* of the accelerating force of gravity was established. The fact, that the higher the point from which a body falls, the greater is the velocity acquired, is patent to observation, though, if we analyse the process by which we arrive at the conclusion, it is by the Method of Concomitant Variations. The *rate* of acceleration, however, is a very difficult and delicate problem to solve. By means of the oscillating pendulum or Atwood's machine (which it is unnecessary to describe here) it is shown (1) that gravity is an uniformly accelerating force, that is, that the increments of velocity in equal times are equal; (2) that the rate of increase varies slightly at different places on the earth's surface; (3) that, in the latitude of Greenwich, in vacuo, and at high-water mark, the rate of acceleration for every second of time is 32·19 inches, the space traversed in the first second of time, if the body fall from rest, being half that quantity, so that the spaces traversed in successive units of time vary as the odd numbers 1, 3, 5 . . . . . $(2n-1)$. A slight degree of attention will show that it is by the Method of Concomitant Variations that all these conclusions are obtained[53].

The conclusions based on statistics in moral and social enquiries are also instances of this application of the Method of Concomitant Variations. It is argued that, if the same causes continue to operate with like intensity and no new causes intervene, the numerical relations established between two classes of social phenomena, as, for instance, deficient education and crime, may be expected to remain constant.

A very important application of the Method of Concomitant Variations is what is now commonly known as the *Historical Method*. The Method designated by this name is, in fact, simply the Method of Concomitant Variations applied to facts supplied by history or to a record of observations on the same classes of facts as those with which history deals. It is specially applicable to those sciences which deal with man as a progressive

[53] The student who wishes for more detailed information on this subject is referred to Professor Price's *Infinitesimal Calculus*, vol. iii. chap. viii. sect. 3.

being, or, at least, a being capable of progress. Thus, a certain institution, custom, or opinion is traced throughout various stages of society, and its growth or decline is connected with that of some other institution, custom, or opinion, or with the general state of civilisation prevalent throughout these periods, it being argued, in the latter case, that, as civilisation advances, the institution, custom, or opinion has grown or declined, as the case may be. This method has of late years been employed with great success in the domains of law, morals, religion, art, and language[54]; and it is sufficient to refer the student for examples to works such as those of Sir Henry Maine, Sir John Lubbock, Professor Max Müller, and Dr. Tylor. The gradual process by which the organisation of the family passes into that of the state, or by which the primitive feeling of resentment is developed into that strict sense of justice which distinguishes civilised man would be amongst the many striking illustrations of this method which are afforded by writers on morals and society, both ancient and modern. When the method is combined with deductions from the science of Psychology, stating *a priori* what might be expected from a general knowledge of human nature, it is called by Mr. Mill the *Inverse Deductive Method*. Under this head I shall briefly advert to it again, in the chapter on the Relation of Induction to Deduction[55].

[54] The instances given on pp. 200–202 are examples of the application of the Historical Method to the Science of Language.

[55] There is one objection to the employment of the Historical Method, which at least demands an answer. The progress, say, of

The Method of Concomitant Variations, especially when applied to subjects other than physical, such as law, morals, language, social statistics, &c., is often called vaguely the *Comparative Method*.

morality, art, or some particular institution, is compared with the progress of general civilisation. But perhaps this very circumstance is amongst the most important considerations to be taken into account in estimating the stage of civilisation to which any people or class has attained. The scientific enquirer, therefore, who employs the Historical Method seems to be open to the objection that he is making one quality vary as an aggregate of qualities of which it is itself one; for, supposing the extreme case of the other qualities which make up the aggregate being all constant, we should then have the identical proposition that the quality or institution in question varies as itself. But, as a matter of fact, we know that the other qualities which make up the aggregate of circumstances which we call civilisation are far from being constant. Moreover, they are all so intertwined with one another that almost any one of them varies directly as almost any other. Amongst these various circumstances, however, we are able to detect one on which all the others seem to be specially dependent. This is, to state it in the most general terms, the intellectual condition prevalent at the given time, in the given place, or amongst the given class. ˙Not only do we find, as a matter of fact, that the current intellectual beliefs and the degree of development of the intellectual faculties is the best index to the state of the other constituents which make up civilisation, but also we should expect *a priori* that these latter circumstances would be mainly determined by the former. For it is by the exercise of reason that man learns, in an infinite variety of ways, to adapt himself to the various circumstances which surround him, that he discovers the means·of gratifying his higher tastes, and that he is enabled to enter into the feelings and understand the wants of others. On this relation between the state of the intellectual faculties and the aggregate of circumstances which constitute civilisation, the student may consult Mr. Mill's *Logic*, Bk. VI. ch. x. § 7.

The term 'Historical,' as designating this Method, is somewhat

Briefly to review these Methods, it will be seen that we can only arrive at absolute certainty by means of one or other of the Methods of Difference, Residues, or Concomitant Variations, while the Method of Agreement and the Joint Method of Agreement and Difference give conclusions only of more or less probability, a probability, however, which sometimes amounts to moral certainty. The Joint Method of Agreement and Difference, or the Double Method of Agreement, possesses one advantage over all the other Methods, namely that, supposing it to have been satisfactorily ascertained by this Method that A is the cause of a, it will follow that it is the *only* cause.

It should also be borne in mind that a wide distinction exists between those cases in which the induction indicates the precise character of the causal connexion which subsists between two or more phenomena and those in which it simply points out that there exists a causal connexion of some kind or other. In the latter case a new induction is required in order to show what the nature of the causal connexion is.

It may be noticed, finally, that the Inductive Methods are strictly reducible to two only, the Method of Agreement and the Method of Difference; the Joint Method of Agreement and Difference being a double employment of the Method of Agreement, supplemented by an employ-

misleading, because, though the facts to which the Method is applied may be mainly supplied by history, they are also, to a large extent, taken from the contemporary observation of tribes and peoples living in different stages, or different phases, of civilisation with reference to the matters of enquiry.

ment of the Method of Difference, the Method of Concomitant Variations being a series of employments of the Method of Difference, and the Method of Residues, though employed in an inductive enquiry, being rather of the nature of a deductive than an inductive method.

Note I.—In the preceding chapter no allusion, or only a casual one, has been made to a circumstance which frequently occasions an insuperable difficulty in the application of the Inductive Methods, namely, the Intermixture of Effects. It has been supposed that the antecedents A, B, C, D, &c. are followed by the consequents a, β, γ, δ, ϵ, &c., the effects being regarded as heterogeneous and not homogeneous. But, suppose the effect of A to be a, of B to be $-\dfrac{a}{2}$, of C to be γ, of D to be $\dfrac{\gamma}{3}$, and of E to be $-\dfrac{\gamma}{2}$, the total effect of A, B, C, D, E will be $\dfrac{a}{2} + \dfrac{5\gamma}{6}$. It is obvious how difficult it would be in this case to discover either the exact portion of the effect which is due to each cause or the several causes which operate to produce the total effect. We might have, in fact, as in mechanical action and reaction, A producing a and B producing $-a$, each cause thus neutralising the effect of the other, so that we might entertain no suspicion that the causes A and B were in operation at all. In these cases, our main resource is Deduction. Having ascertained separately by one or other of the various inductive methods, or from previous deductions, the effects, say of A, B, C D, we calculate deductively their combined effect, and then, by subtracting, according to the Method of Residues, the sum of the known causes from the total aggregate of causes and the known portion of the effect from the total effect, we simplify, if we do not solve, the problem. On the insufficiency, under ordinary circumstances, of the Inductive Methods, without the aid of Deduction, to grapple with cases of this kind[56], and on the nature of the assistance rendered by Deduction, the reader may consult Mr. Mill's *Logic*, Bk. III. ch. x. § 4–8, and ch. xi.

In cases of this kind, where the action of one cause is augmented, diminished, or wholly counteracted by that of another, it must not be supposed that any part of its appropriate effect has failed to be produced, even though it may have disappeared wholly or partially in the total

[56] Since the appearance of the first edition of this work, it has been pointed out by Mr. Bain that 'Concomitant Variation is the only one of the [Inductive] Methods that can operate to advantage in such cases.' I take the liberty of transcribing the passage: 'If a cause happens to vary alone, the effect will also vary alone, and cause and effect may be thus singled out under the greatest complications. Thus, when the appetite for food increases with the cold, we have a strong evidence of connexion between those two facts, although other circumstances may operate in the same direction.

'The assigning of the respective parts of the sun and moon, in the action of the Tides, may be effected, to a certain degree of exactness, by the variation of the amount according to the positions of the two attracting bodies.

'By a series of experiments of Concomitant Variations, directed to ascertain the elimination of nitrogen in the human body under varieties of muscular exercise, Dr. Parkes obtained the remarkable conclusion, that a muscle grows during exercise and loses bulk during the subsequent rest.'—Bain's *Logic*, Bk. III. ch. viii. § 6.

P

result. An object may remain at rest, when subject
to two equal forces acting in opposite directions, but
we cannot say of either of these forces that it is inoperative : each, it is true, prevents any visible effect
resulting from the other ; but then this is the very effect
which it produces, and the correct mode of describing
either of the opposing forces would be to say that
it has a *tendency* to make the given object move with
a certain velocity in a certain direction. The student
cannot too constantly bear in mind that every cause
invariably produces its full effect, though other causes
may prevent that effect from manifesting itself with all
the intensity with which it would manifest itself, if it acted
alone ; that there are, strictly speaking, no exceptions to
laws of nature, though these laws, in their manifold action
and reaction, may modify or even neutralise each other.
The aphorism 'Every rule has an exception,' is only true,
even in Grammar, either because the rule is inexactly
stated or because it conflicts with some other rule known
or unknown.

Note 2.—The Canons for the Inductive Methods were
first stated by Mr. Mill, and the importance now attached
to them in most analyses of inductive enquiries is mainly
due to his influence. The methods are, however, as
Mr. Mill himself states, 'distinctly recognised' in Sir
John Herschel's *Discourse on the Study of Natural Philosophy*, so often quoted in this work, 'though not so
clearly characterised and defined, nor their correlation
so fully shown, as has appeared to me desirable.' In

the Second Book of Bacon's *Novum Organum*, we find some approximations, very rough, it is true, to formal inductive methods. The 'instantiæ crucis' have already been adduced as examples of the Method of Difference, and the 'instantiæ solitariæ' as comprising examples of both the Method of Agreement and the Method of Difference; but the part of the *Novum Organum* to which I am now alluding, and which is intended to be of more universal application than the 'instantiæ crucis' and the 'instantiæ solitariæ,' is contained in the early Aphorisms of the Second Book. Certain Tables of Instances are there given for the purpose of providing materials with which to conduct an investigation into what Bacon called the 'Form [57],' corresponding pretty nearly, at least in this connexion, with what we should call the 'Cause,' of Heat. The instances are very far from satisfying the conditions of Mr. Mill's Methods, but the principles on which they are arranged in Tables bear a close analogy to the principles on which the Canons are constructed. The best mode, perhaps, of enabling the student to perceive the extent of the resemblance is to state the conditions with which the instances in Bacon's Tables would be required to conform, in order to satisfy the requirements of Mr. Mill's Methods.

If the 'Instantiæ convenientes in natura calidi' [58] were

---

[57] On the meaning attached by Bacon to the word 'Form,' and its relation to 'Essence' and 'Cause,' see the Introduction to my Edition of the *Novum Organum*, § 8.

[58] *Novum Organum*, Lib. II. Aph. xi.

so related to one another that, besides the given phenomenon (heat), only one other circumstance were common
to them all, that other circumstance might be regarded,
with more or less probability, as the cause (or effect) of
heat, or, at least, as connected with it through some fact
of causation. Such instances would then come under
the Method of Agreement.

If one instance in the Table of Agreement ('Instantiæ
convenientes in natura calidi') were so related to one
of the instances in the Table of Privation ('Instantiæ in
proximo, quæ privantur natura calidi')[59] as to have every
circumstance in common with it, except that the former,
besides presenting the phenomenon of heat which is
supposed to be absent in the latter, also presented some
other circumstance which was absent from the latter, this
other circumstance would be the cause (or effect), or a
necessary part of the cause, of heat. We should here
have the Method of Difference.

If, in the 'Tabula graduum, sive comparativæ in
calido[60],' we could discover some one phenomenon which
increased and diminished proportionately with the increase
and diminution of heat, that phenomenon would be the
cause or the effect of heat, or, at least, connected with it
through some fact of causation, and the conditions would
thus conform with the requirements of the Method of
Concomitant Variations. If it could be shown that this
phenomenon and heat were the *only* circumstances
which varied concurrently, then the phenomenon would

[59] *Novum Organum*, Lib. II. Aph. xxii.　　　[60] Id. Aph. xiii.

be proved to be either the cause or the effect of heat, and would conform with the requirements of the rider to this last Method (p. 186).

The 'Exemplum exclusivæ, sive rejectionis naturarum a forma calidi'[61] (which is based on the foregoing Tables) bears some, though, it must be acknowledged, a very slight, resemblance to the Method of Residues. These 'rejectiones' consist in excluding some possible explanation of the phenomenon, either because an instance, which does not present the phenomenon, does present the assigned cause, or because an instance, which does present the phenomenon, does not present the assigned cause[62] (and similarly with regard to increase and decrease). As an example of the former case, we may take the following 'rejectio': 'Per radios lunæ (which were then supposed to be cold) et aliarum stellarum rejice lucem et lumen.' As examples of the latter, we may take the two following: 'Per radios solis, rejice naturam elementarem (that is, 'terrestrial nature,' which is composed of 'the four elements'); Per ignem communem, et maxime per ignes subterraneos (qui remotissimi sunt, et plurimum intercluduntur a radiis cœlestibus) rejice naturam cœlestem.' By a succession of these 'rejectiones,' we limit the number of possible explanations, amongst which we are to look for the true one. Bacon's 'rejec-

[61] *Novum Organum*, Lib. II. Aph. xviii.

[62] The latter, of course, is not a legitimate argument. The effect may be due to several distinct causes, a fact which was not recognised by Bacon. See my notes on *Novum Organum*, Lib. II. Aph. xvi.

tions,' however, not being, as a matter of fact, exhaustive, lead to a purely negative result; they may save us from unnecessary trouble in seeking for a cause where it cannot be found, but they do not, like the Method of Residues, leave a definite number of antecedents which either constitute the cause, or amongst which we know that the cause is to be sought.

It is plain that if there were a certain number only of possible causes of the given phenomenon, and by the method of rejections we could exclude all but one, this one remaining cause would be the undoubted cause of the given phenomenon. This case Bacon appears to have regarded as the perfect type of Induction, and as alone capable of affording certainty [63].

Note 3.—Dr. Whewell (in a pamphlet published in 1849, which is now embodied in the *Philosophy of Discovery* [64]) questions the utility of the Four Methods.

---

[63] It must be understood that, in this note, I am simply comparing the 'Tables' of Bacon with the 'Methods' of Mr. Mill. On the relation of the 'Tables' to each other and on the special importance attached by Bacon to the 'Rejections,' the student may consult § 9 of the Introduction to my edition of the *Novum Organum* (Clarendon Press) and my notes to the earlier aphorisms of the Second Book.

In comparing the logical procedure of Bacon and Mill, it should be carefully borne in mind that Bacon contemplated the concurrent use of all the Tables, as preparatory to his Method of Rejections, and regarded the construction of the Tables and the subsequent application to them of the Method of Rejections as constituting only one process. On the other hand, each of Mr. Mill's Methods may be worked independently, and lead to a final conclusion.

[64] See *Philosophy of Discovery*, ch. xxii. The criticism of Mr.

'Upon these methods,' he says, 'the obvious thing to remark is, that they take for granted the very thing which is most difficult to discover, the reduction of the pheno- mena to formulæ such as are here presented to us.' He also objects that, as a matter of fact, no discoveries have ever been made by the employment of these methods. 'Who will carry these formulæ through the history of the sciences, as they have really grown up, and show us that these four methods have been operative in their formation; or that any light is thrown upon the steps of their progress by reference to these formulæ?'

The first objection is, as Mr. Mill points out, of the same character with the objections raised by Locke and other writers of the eighteenth century against the Rules of Syllogistic Reasoning. The reply, in either case, is that Logic does not profess to *supply* arguments, but to *test* them. Men have certainly reasoned, and reasoned with the greatest success, without any conscious use of the rules of Logic. But it is the province of a system of Logic to analyse the arguments commonly employed, to discriminate between those which are correct and those which are incorrect, and thus to enable men to detect, in the case of others, and to avoid, in their own case, erroneous methods of reasoning. To think of appro- priate arguments is undoubtedly more difficult than to test them; but this fact does not obviate the necessity of submitting them to a test. Nor is it a more real objec-

Mill's Methods will be found in §§ 38–48. Mr. Mill replies in a note at the end of Bk. III. ch. ix.

tion that men, who know nothing of the technical rules of Logic, often reason faultlessly themselves, and show remarkable acuteness in detecting inconclusive reasoning in the arguments of others. Many men speak grammatically without having learnt any system of grammar; in the same manner, many men reason logically without having learnt any system of Logic. But the great majority of men, there can be little doubt, may derive assistance both from one and the other. Grammar fulfils its functions when it raises the student to the level of the most correct speakers; similarly, Logic fulfils its functions when it raises the student to the level of the best reasoners. As applied to the syllogistic rules and formulæ, this defence would now be generally admitted, but it holds equally good of the methods under which it may be shown that our inductive arguments may ultimately be arranged. 'The business of Inductive Logic,' says Mr. Mill, 'is to provide rules and models (such as the Syllogism and its rules are for ratiocination) to which if inductive arguments conform, those arguments are conclusive, and not otherwise. This is what the Four Methods profess to be, and what I believe they are universally considered to be by experimental philosophers, who had practised all of them long before any one sought to reduce the practice to theory.'

With regard to the second objection, that these methods have not been operative in the formation of the sciences, Dr. Whewell seems to ignore the distinction between the conscious and the unconscious employment of a method. It is undoubtedly true that in records of scientific investigations we seldom find the formal language in which the Inductive Canons are expressed. It seems to me equally true that in such records we invariably detect the employment of the Canons themselves. Discoveries are of two kinds : they are either entirely the result of patient research, or they are first suggested to the mind by some brilliant thought, and afterwards verified by rigorous proof. In the former case, the discoverer must have made sure of his ground as he proceeded, and, so far as his method was inductive, he could only do so by appealing, consciously or unconsciously, to one or more of the inductive methods; if he acted otherwise, he arrived at a true result by mere accident. In discussing the latter case, I must repeat what has already been stated, that it is not the office of Logic, either inductive or deductive, to *suggest* thoughts, but to *analyse* and to *test* them. Now, in the case we are supposing, the discovery really consists of two parts —the original conception and the subsequent process by which it is determined to be the true explanation of the phenomenon. However striking and appropriate the conception, we have no right to regard it as the true explanation of the phenomenon till it has been subjected to the most rigorous investigation. This investigation must be either inductive or deductive, or both. But, so far as it is inductive, it must conform with the requirements of the Inductive Canons, or else it will not result in positive proof, or even approximate closely to it. As in the former case, unless the discoverer has, consciously or unconsciously, reasoned in strict conformity with the requirements of Logic, he has no right to feel any confidence in the result of his researches.

It may be added that appropriate conceptions, promising to be fertile in scientific results, are only likely, as a rule, to occur to persons whose minds have been habitually disciplined by the strict observance, conscious or unconscious, of the laws of reasoning. Originality is not a quality, as some seem to think, which admits of no psychological explanation.

I have not thought it desirable to discuss more recent criticisms of the Inductive Methods, because, apart from the stress which they lay on the difficulty of satisfying the conditions of the Canons (a difficulty which is acknowledged on all sides to exist, at least in many cases), I cannot think that they have added materially to the objections raised by Dr. Whewell.

The student is particularly requested to read, in connexion with this chapter, the 'Preface to the Third Edition,' reprinted at the beginning of the Book. This Preface deals with certain controverted points respecting the certainty of Inductive Reasoning and the nature of the assumptions made in it, with which, though they could not conveniently be introduced into the body of the book, it is desirable that the student should acquaint himself.

CHAPTER IV.

Of Imperfect Inductions.

AN argument from the particular to the general, or from particulars to adjacent particulars, may fall short of absolute proof, or even of moral certainty, while it commends itself as possessing more or less of probability. Arguments of this character may be called Imperfect Inductions. Under this head fall imperfect applications of the experimental or inductive methods, the argument from analogy, and incomplete cases of *Inductio per enumerationem simplicem.*

The *Inductio per enumerationem simplicem* is, as already noticed[1], when *complete* [*Inductio Completa*], a deductive, and not an inductive, argument. When *incomplete*, it is an inductive argument, for it is an inference of the general from the particular or the unknown from the known. This form of Induction affords certainty only when, as in the case of the Laws of Universal Causation and of the Uniformity of Nature, or of the Mathematical Axioms,

[1] See p. 125, note 2, and *Deductive Logic*, Part III. chap. i. appended note 2.

it is grounded upon universal experience, and we feel assured that, if there had been at any time or were now in any place any instance to the contrary, it would not have escaped our notice. But, in ordinary cases, the incomplete *Inductio per enumerationem simplicem* affords only a presumption, sometimes very slight, sometimes tolerably strong, in favour of the position which it is adduced to establish. I perceive, say in five, ten, or twenty cases, that the phenomenon a is attended by the phenomenon b, and, knowing of no cases in which the one phenomenon is not attended by the other, I begin to suspect that a and b are connected together in the way of causation. Such a surmise may afterwards be proved by the aid of one or other of the five methods to be correct, and, in that case, it is taken out of the category of inductions *per enumerationem simplicem*, and becomes an instance of a scientific induction. But, if neither proved nor disproved, it still has a certain amount of probability in its favour, that amount depending on the two following considerations: (1) the number of positive instances which have occurred to us; (2) the likelihood, if there be any negative instances, of our having met with them. The first of these considerations deserves little weight, unless supported by the other. A native of the North of Europe, some centuries ago, might, if the mere accumulation of positive instances were sufficient, have taken it for a certain truth that all men had white complexions. His own personal observation, as well as the reports of travellers and the traditions of his race, would have furnished numberless instances in favour of the position. But, before drawing the in-. ference, he ought to have reflected that he possessed information about a small portion only of the inhabitants of the earth's surface, that a difference of climate might produce a difference of complexion, and that there was no reason for supposing that the anatomical structure of man, or the various characteristics which we denominate human, are necessarily connected with a skin of one particular colour. But, on the other hand, we may affirm with tolerable certainty that all the varieties of beings possessing the physical structure of man have the capacity of articulate speech; for, if there were any races exhibiting the one set of phenomena without the other, there is every probability, with our present knowledge of the earth's surface, that we should be acquainted with their existence. In this instance the first consideration, which in itself would deserve little weight, is converted into a certainty almost absolute by the support which it derives from the second.

It cannot be too strongly impressed on the mind of the student that a mere *enumeratio simplex*, that is, a mere assemblage of positive instances, unless we have reason to suppose that, were there any instances to the contrary, they would have become known to us, is simply worthless. '*Inductio quæ procedit per enumerationem simplicem* res puerilis est.' But if the *enumeratio simplex* be accompanied by a well-grounded conviction that there are no instances to the contrary, it may afford a very high degree of probability, and, if we can assure our-selves that there are no instances to the contrary, to us individually it will afford certainty.

It might seem that an Inductio per Enumerationem Simplicem is always an employment of the Method of Agreement. But there is this essential difference. The Method of Agreement is a method of *elimination*, select-ing some and rejecting other instances, and founding its conclusion not on the quantity but on the *character* of the instances which it selects. The Inductio per Enu-merationem Simplicem, on the other hand, depends for its validity on the *number* of instances; the instances, indeed, must be gathered from every available field, and hence sometimes we speak of their variety as well as their quantity, but the one essential characteristic of the method is that it does not select, but *accumulate* instances. A few well-selected instances are often suffi-cient to satisfy the requirements of the Method of Agreement. The same number, when we abstract the grounds on which they were selected, would be utterly insufficient to justify an Inductio per Enumerationem Simplicem.

It may in fact be remarked of all the Experimental Methods that they are devices for saving labour. The range of our experience is often insufficient to justify an argument founded on an Inductio per Enumerationem Simplicem, but by means of the Experimental or In-ductive Methods we so *select* our instances as to bring the particular case which we are investigating under the general laws of Universal Causation and the Uniformity of Nature. The validity of the induction in question is thus artificially connected with the validity of these universally accepted inductions, and we are enabled to argue from the truth of the latter to that of the former.

Uncontradicted experience, of course, implies a great variety of instances, and, from this point of view, every well-grounded Inductio per Enumerationem Simplicem might be represented as an application of the Method of Agreement. But to represent it in this form would often weaken its force. For, while our experience may be so wide as to justify us in affirming the constant union of two or more circumstances, the number of other common circumstances, known or suspected, with which these are found in invariable combination, may be so large as to render it impossible for us to satisfy even approximately the conditions of the Method of Agreement. Here, as elsewhere, an argument often admits of being stated in two ways, and it is the office of the logician to state it in that form in which it carries the largest amount of conviction, or rather offers the most satisfactory kind of proof.

It is, as I have already pointed out in the First Chapter [2], by means of an Inductio per Enumerationem Simplicem that we establish what have been called 'Inductions of Co-existence.' This is the case, when, as the result of a wide experience, two phenomena

[2] Pp. 7–9.

are found to be invariably co-existent, but we have no evidence to connect them as cause and effect, or even as effects of the same cause. Such are the attributes which are found to be invariably united in the same Natural Kinds, that is to say, in the same species of plants, animals, and minerals; such are the two properties of Inertia and Gravity which are found united in all matter. In all these cases, there is probably some causal connexion, hitherto undetected, between the co-existing phenomena; but while we are unable to apply with any success the more refined inductive methods, we must content ourselves with regarding the uniformity as simply one of co-existence. If we made any progress towards the discovery of a causal connexion, the uniformity would be transferred to another category, and would rank amongst the inductions discussed in the last chapter. Meanwhile, these inductions, depending simply on uncontradicted experience, and being at present inaccessible to the Methods of Elimination, must be regarded as generalisations awaiting further investigation [3].

The term 'Empirical Generalisation' or 'Empirical

---

[3] For a further discussion of the Uniformities of Co-existence, the reader is referred to Mr. Bain's *Logic*, Bk. iii. ch. 3. I am disposed to estimate more highly than Mr. Bain the probability that these uniformities might, if our knowledge were extended, be ultimately resolved into Uniformities of Causation, and hence they do not appear to me to require any separate or detailed treatment in a work on Logic.

Law' might be conveniently appropriated to express those secondary laws (as distinct from Ultimate Laws of Nature[4]) which are the result of an Inductio per Enumerationem Simplicem. Though these expressions are employed with great latitude, it is usually regarded as characteristic of an Empirical Law or Generalisation that it can only be received as true within the limits of the data from which it is derived, that at another time, at another place, or under different circumstances from those under which the observations were made, it might be found to break down[5]. It is true that, owing to the conflict of causes, this description applies to many of the conclusions arrived at by means of the Inductive

[4] Some of these Ultimate Laws of Nature, such as the Law of Universal Causation, the Law of the Conservation of Energy, the invariable co-existence of Inertia with Gravity, &c., appear to rest simply on uncontradicted experience, that is to say, on an Inductio per Enumerationem Simplicem, and still it would seem paradoxical to speak of them as merely 'Empirical Laws.' An Empirical Law might, perhaps, be defined as a secondary law, the causal derivation of which is not yet known or even surmised with any probability, or as a subordinate generalisation arrived at by an Inductio per Enumerationem Simplicem; definitions which, it will be perceived, are really identical. I have, however, avoided any special discussion of what are called 'Empirical Laws,' both on account of the extremely indeterminate use of the expression, and because such a discussion is calculated, in my opinion, needlessly to perplex the student by the complicated questions to which it leads. The advanced student can refer to Mr. Mill's *Logic*, Bk. III. ch. xv., and Bk. V. ch. v. § 4, but he will be introduced, I venture to suggest, to more difficulties than he will find solved.

[5] See Herschel's *Discourse on the Study of Natural Philosophy*, § 187, and Mill's *Logic*, Bk. III. ch. xvi. § 4.

Methods, but it is peculiarly applicable to the results of the Inductio per Enumerationem Simplicem, and it would be extremely convenient to possess an expression by which the results of this method might be at once distinguished from those of scientific induction on the one hand, and those of analogy (to be discussed presently) on the other. Instances of Empirical Laws in this restricted sense are such generalisations as that certain animals or flowers[6] are of a certain colour, that certain tribes of men are less capable of civilisation than others, and, perhaps, that certain appearances of sky are indicative of certain changes of weather. There are, of course, some cases in which it is difficult to determine whether a given conclusion has been arrived at by the Inductio per Enumerationem Simplicem or by an imperfect application of the Method of Agreement, that is to say, whether it is based on instances taken indifferently, or on selected instances.

Another form of imperfect induction is the Argument from Analogy[7]. Here we do not argue from a number

[6] The colours of flowers, however, seem to be in a fair way of being accounted for by the peculiarities of their mode of fertilisation. See a most interesting work on the Colours of Flowers, by Mr. Grant Allen, published in Macmillans' *Nature Series*, 1882.

[7] It will be observed that the word 'Analogy' is here employed in the sense of 'resemblance.' In the stricter and more ancient meaning of the term, it signifies an equality of relations (ἰσότης λόγων). See Aristotle's *Ethics*, Bk. v. 3 (8). The reader will find the two significations of the word 'Analogy' discriminated in the *Elements of Deductive Logic*, Part III. ch. i. note 2.

of instances, as in the case of Inductio per Enumerationem Simplicem, but from a number of points of resemblance. The argument is not, that, because S, T, U, V, W, &c. exhibit the union of *m* with *a*, *b*, *c*, we may therefore expect to find *m* in Z, or wherever else *a*, *b*, *c* may occur; but that, because X and Y (any two or more instances) agree in the possession of certain qualities *a*, *b*, *c*, we may expect to find the quality *m* which is presented by X exhibited also in Y. The argument is based, not on the number of *instances* in which the two sets of qualities are found united, but on the number of *qualities* which are found to be common to two or more instances: the argument is not that I have so often observed *a*, *b*, *c* in conjunction with *m*

Archbishop Whately defines Analogy as a resemblance of Relations. This definition, if intended to represent the ancient signification of the word, is incorrect. The Aristotelian Analogy is an *equality*, not a *resemblance* of relations. The instance given in *Eth. Nic.* i. 6 (12) is that, in man, the reason ($\nu o \hat{v} s$) bears to the living principle ($\psi v \chi \acute{\eta}$) the same relation that the faculty of vision ($\ddot{o} \psi \iota s$) bears to the body ($\sigma \hat{\omega} \mu a$): $\dot{\omega} s \gamma \dot{a} \rho \dot{\epsilon} \nu \sigma \acute{\omega} \mu a \tau \iota \ \ddot{o} \psi \iota s, \dot{\epsilon} \nu \psi v \chi \hat{\eta} \ \nu o \hat{v} s.$ The assertion, in this instance, it will be noticed, is that the relation to each other of the two former members of the analogy is, not *similar* to, but the *same* as, that of the two latter. The Aristotelian term $\dot{a} \nu a \lambda o \gamma \acute{\iota} a$, in fact, exactly corresponds with the term Proportion as employed by mathematicians, and it was by the word Proportio, when not availing themselves of the Greek word Analogia itself, that the Romans expressed this form of argument. See Quinctilian, *Inst. Orat.* i. 6: '*Analogiæ* quam proxime ex Græco transferentes in Latinum *proportionem* vocaverunt, hæc vis est: Ut id, quod dubium est, ad aliquid simile, de quo non quæritur, referat; ut incerta certis probet.' I am indebted for this quotation to Mr. Austin's *Lectures on Jurisprudence*, vol. iii. p. 255.

that I believe these qualities to be conjoined invariably, but that I know X and Y to resemble each other in so many points that I believe them to resemble each other in all.

Thus, because the moon resembles the earth in being a large spheroid revolving round another body, as well as in various other particulars, it may be argued that it probably resembles the earth also in sustaining animal and vegetable life on its surface. But, if every ground of resemblance furnishes a probable reason for assigning to the one body any property known to belong to the other, it is evident that every ground of dissimilarity will also furnish a probable reason for denying of the first body any property known to belong to the second. In estimating, therefore, the value of an analogical argument, we must strike a balance between the known points of resemblance and the known points of difference, and according as the one or the other preponderate, and in the proportion in which the one or the other preponderate, is the weight of the argument to be regarded as inclining. If, for instance, the phenomenon A is known to resemble the phenomenon B in four points, whereas the known points of difference between them are three, and it is discovered that some new property belongs to A but it is uncertain whether it also belongs to B, the value of the analogical argument that it does belong to B will be represented by 4 : 3.

Before, however, we are justified in drawing this inference, it is necessary to observe certain cautions.

In the first place, we must have no evidence that there is any causal connexion between the new property and any of the known points of resemblance or difference. If we have such evidence, the argument ceases to be analogical, and, if not a perfect induction, is an imperfect induction of the kind to be described presently. We know, for instance, that animal and vegetable life on the surface of the earth could not exist without moisture; but, so far as we are able to ascertain, there is no moisture on the surface of the moon. Hence we appear to be justified in concluding, not by analogy, but by the Method of Difference (assuming, of course, the accuracy of the observations), that animal and vegetable life, in the sense ordinarily attached to those terms, are not to be found on the moon's surface [8]. Again, we happen to know two men who bear a considerable resemblance to each other in character and opinions. One of these men acts in a particular way, and we infer, analogically, that the other will act similarly. But, suppose we ascertain that the act of the former man was due to some particular characteristic, say avarice. The inference will now no longer depend on the ratio of the known points of resemblance to the known points of difference in the characters and opinions of the two men, that is, on analogy, but it will depend mainly on the presence or absence, the strength or weak-

[8] See the essay *Of the Plurality of Worlds* (usually attributed to Dr. Whewell), ch. ix. sect. 7–9. The whole of this essay furnishes excellent examples of the employment of the Argument from Analogy, and also illustrates the extreme caution and delicacy which are requisite in estimating its value.

ness, of this particular characteristic in the second man, and, in a subsidiary degree, on the presence or absence, the strength or weakness, of corroborating or countervailing motives ; that is, it will depend, not on analogy, but on other modes of induction.

Secondly, though there must be no evidence to connect the property in question with any of the known points of resemblance or difference, there must, on the other hand, be no evidence to disconnect it. If there be such evidence, the point of resemblance or difference with which we know or believe it to be unconnected must, in estimating the value of the analogy, be left out of consideration. The reason is obvious. When we are enquiring whether this property is more likely to be connected with the known points of resemblance or the known points of difference, it is plain that we must only take into account those points with which there is, at least, some chance of its being connected.

Thirdly, we must have no reason to suspect that any of the known points of resemblance or difference, of which the argument takes account, are causally connected with each other. If the compared phenomena agree in the possession of the properties a, b, c, d, e, and of these properties b is an effect of (or causally connected with) a, and d is an effect of (or causally connected with) c, the only properties which ought to be taken into account in estimating the value of the analogy are a, c, e. The moon is supposed to differ from the earth in having no clouds and no water, but, as these two properties are mutually connected in the way of cause and effect, they can only be allowed to count as one item in instituting a comparison, for the purposes of analogy, between the known points of resemblance and the known points of difference in the two bodies. The enormous difference, on the other hand, between the maximum and minimum temperature of any place on the moon's surface, owing to the extreme length of the lunar days and nights and the absence of any sensible atmosphere, constitutes a distinct point of difference, and, as such, furnishes an additional argument against the habitation of the moon. When we ask to which side the argument from analogy inclines, we are asking whether it is more probable that the property in question (known to belong to the one phenomenon, but not known either to belong or not to belong to the other) is connected, by way of causation, with one of the known points of resemblance, or with one of the known points of difference : but, in calculating the probability, it is essential that every point should, so far as we know, be independent of every other ; for it is only in virtue of each being supposed to be an ultimate property or to point to an ultimate property that it has any claim to be taken into the account. Thus, if any two of the properties are found to be joint effects of the same cause or to stand to each other in the relation of cause and effect, they furnish only one argument instead of two. If we say of A that he is likely, under some particular conjuncture of circumstances, to act in the same manner as

B, because they are both of them vain and selfish, we shall not strengthen our argument by adding a number of characteristics which are deducible from vanity.and selfishness, or by adducing a number of individual acts in which these qualities have been exhibited.

Fourthly, it is only when we have reason to suppose that we are acquainted with a considerable proportion of the properties of two objects, that the argument from analogy can have much weight. If we know only a few properties out of a large number, they may happen to be precisely those which are exceptional rather than representative, points of similarity where the objects themselves are mainly dissimilar, or points of dissimilarity where the objects are mainly similar. Thus, we know that in some respects the planet Mars closely resembles the earth, as, for instance, in having an atmosphere, a surface distributed into land and water, and probably a temperature in which life similar to that on our own globe might exist: but it would be very rash to conclude from these data that it also resembles the earth in sustaining animal and vegetable life on its surface; for, though life, such as we understand it, does not appear to be impossible on the planet Mars as it appears to be on many of the other celestial bodies, the number of properties with which we are acquainted is so small as compared with the number of properties with which we are unacquainted that there is little or nothing on which to ground even a probable conclusion. On the other hand, the analogy by which Kepler boldly extended the three laws gained from the observation of the motion of Mars to the remaining planets was a perfectly sound one; for the orbit of a planet, as compared with the condition of its surface, is a very simple phenomenon, and what was known of the orbits of the other planets made it appear more likely that they would correspond with the orbit of Mars than that they would differ from it.

The value of the Argument from Analogy, then, we see, depends on the ratio of the ascertained points of resemblance to (1) the ascertained points of difference, (2) the entire assemblage of the properties of the objects compared. If the ascertained resemblances are numerous, the ascertained differences few, and we have reason to think that we are well acquainted with the objects compared, the argument from analogy is very forcible. If, on the other hand, the ascertained resemblances only slightly exceed in number the ascertained differences, or if we have reason to suppose that there are numerous properties in the compared objects with which we are unacquainted, the value of the argument from analogy may be very slight. It is commonly said that the value of an argument from analogy ranges from certainty to zero. If it reaches certainty, the argument becomes a complete induction; if it falls to zero, it ceases to be an argument at all; if the probability is expressed by less than one-half, that is, if the number of ascertained resemblances be less than the number of ascertained differences, it is usual to say that analogy

is against the possession by the one object of a quality
known to belong to the other, or, in other words, in
favour of their differing in the possession of this quality
rather than agreeing in it.

'Besides the competition between analogy and diversity,' says Mr. Mill [9], 'there may be a competition of
conflicting analogies.' An object may be known to
resemble one object in some particulars and another in
others, and it may be a question with which of the two
it ought to be classed, or which of the two it is the more
likely to resemble in some unknown property. Thus,
for some time it was a question whether a sponge was
an animal or a vegetable substance; and it is often by
conflicting analogies that we attempt to determine to
which of two or more masters a painting or a statue
should be ascribed.

The extreme caution which is requisite in employing the
Argument from Analogy may be illustrated by the following scientific errors which have resulted from a hasty and
inconsiderate employment of this mode of reasoning.

Sir W. Grove, in his *Correlation of Physical Forces* [10],
while combating the once fashionable doctrine of electrical fluids, brings into juxta-position two very interesting instances of hasty analogies.

'The progressive stages,' he says, 'in the History of Physical Philosophy will account in a great measure for the
adoption by the early electricians of the theories of fluids.

[9] Mill's *Logic*, Bk. III. ch. xx. § 2.
[10] Fifth edition, p. 135.

'The ancients, when they witnessed a natural phenomenon, removed from ordinary analogies, and unexplained by any mechanical action known to them, referred it to a soul, a spiritual or preternatural power: thus amber and the magnet were supposed by Thales to have a soul; the functions of digestion, assimilation, &c., were supposed by Paracelsus to be effected by a spirit (the Archæus). Air and gases were also at first deemed spiritual, but subsequently became invested with a more material character; and the word gas, from *geist*, a ghost or spirit, affords us an instance of the gradual transmission of a spiritual into a physical conception.

'The establishment by Torricelli of the ponderable character of air and gas, showed that substances which had been deemed spiritual and essentially different from ponderable matters were possessed of its attributes. A less superstitious mode of reasoning ensued, and now aëriform fluids were shown to be analogous in many of their actions to liquids or known fluids. A belief in the existence of other fluids, differing from air as this differed from water, grew up, and, when a new phenomenon presented itself, recourse was had to a hypothetic fluid for explaining the phenomenon and connecting it with others; the mind, once possessed of the idea of a fluid, soon invested it with the necessary powers and properties, and grafted upon it a luxuriant vegetation of imaginary offshoots.'

Most of my readers will be aware of the difficulties experienced by the early geologists in accounting for the fact that the strata of our own and other northern countries often contain remains of animals and shells akin to those which are now to be found only in the torrid zone. This difficulty is easily explained by supposing a different distribution of land and water over the surface of the globe from that which at present exists. But we must pause before we admit the inference that, because these animals and shells are *akin* to those which are now found only in warm climates, they must, therefore, have subsisted in a similar temperature.

'When reasoning on such phenomena,' says Sir Charles Lyell[11], 'the reader must always bear in mind that the fossil individuals belonged to *species* of elephant, rhinoceros, hippopotamus, bear, tiger, and hyæna, distinct from those which now dwell within or near the tropics. Dr. Fleming, in a discussion on this subject, has well remarked that a near resemblance in form and osteological structure is not always followed, in the existing creation, by a similarity of geographical distribution; and we must therefore be on our guard against deciding too confidently, from mere analogy of anatomical structure, respecting the habits and physiological peculiarities of *species* now no more. "The zebra delights to roam over the tropical plains; while the horse can maintain its existence throughout an Iceland winter. The buffalo, like the zebra, prefers a high temperature, and cannot thrive even where the common ox prospers. The musk ox, on the other hand, though nearly resembling the buffalo, prefers the stinted herbage of the arctic regions, and is able, by its periodical migrations, to outlive a northern winter. The jackal (*Canis aureus*) inhabits Africa, the warmer parts of Asia, and Greece; while the isatis (*Canis lagopus*) resides in the arctic regions. The African hare and the polar hare have their geographical distribution expressed in their trivial names;" and different species of bears thrive in tropical, temperate, and arctic latitudes.

'Recent investigations have placed beyond all doubt the

---

[11] Lyell's *Principles of Geology,* ch. vi. (ninth edition); ch. x. (tenth edition).

important fact that a species of tiger, identical with that of Bengal, is common in the neigbourhood of Lake Aral, near Sussac, in the forty-fifth degree of north latitude ; and from time to time this animal is now seen in Siberia, in a latitude as far north as the parallel of Berlin and Hamburgh. Humboldt remarks that the part of Southern Asia now inhabited by this Indian species of tiger is separated from the Himalaya by two great chains of mountains, each covered with perpetual snow,—the chain of Kuenlun, lat. 35° N., and that of Mouztagh, lat. 42°,—so that it is impossible that these animals should merely have made excursions from India, so as to have penetrated in summer to the forty-eighth and fifty-third degrees of north latitude. They must remain all the winter north of the Mouztagh, or Celestial Mountains. The last tiger, killed in 1828, on the Lena, in lat. 52¼°, was in a climate colder than that of Petersburg and Stockholm.'

Neither through Analogy nor through Induction by Simple Enumeration can we establish a fact of Causation, though the conclusions of either of these methods may suggest to us such a fact. When we begin to suspect that any one circumstance or set of circumstances is the cause or the effect of another, or connected with it in the way of causation, we ought at once to attempt to apply, if possible, one or more of the Experimental Methods. If we can satisfy ourselves that their conditions, or those of any one of them, have been rigorously fulfilled, we have, of course, obtained a Valid Induction, giving us either absolute or moral certainty. But something considerably short of a rigorous fulfilment of these conditions may still lead to a conclusion, possessing more or less of probability. We ·may, for instance, to take the Method of Agreement, feel uncertain whether a and b (any two circumstances) are the only material circumstances which the cases we have examined exhibit in common ; but still we may have examined so many, so various, and so well selected instances, that we may be justified in regarding it as highly probable that the two circumstances stand to each other in the relation of cause and effect, or are, at least, connected in the way of causation. Similarly, to take the Method of Difference, in the act of introducing a new antecedent, we may have unwittingly introduced some other new antecedent, or, in omitting an antecedent, we may have unwittingly introduced or omitted some other antecedent; but still we may have exercised such extreme caution as to justify us in feeling an assurance amounting almost, though not altogether, to certainty that the experiment has been rightly performed. The less our assurance of this fact, the slighter is the probability of the conclusion.

There remains one case, which is attended with some perplexity. It sometimes happens that, though we may be unable to establish a fact of causation between two particular phenomena, we may be able to show that some one phenomenon stands in a causal relation to some one or other of a definite number of other phenomena. Thus, supposing a vegetable to be transplanted to a distant part of the world, we may be able to assure ourselves, by excluding other causes of difference, that any new qualities which it may assume are due either to difference of climate, or to difference of soil, or to both these causes conjointly, though our knowledge may not enable us to assign amongst these alternatives the particular cause or combination of causes to which the effect is due. Now ought such an Inference to be classified as a perfect or an imperfect Induction? If we content ourselves with stating the alternatives, the inference should be regarded, so far as it goes, as a Perfect Induction; for within the limits stated the conclusion may be considered absolutely certain. But if, on any grounds, we suppose one of these alternatives to be more probable than the others, and we state this as our conclusion, the inference is, of course, only a probable one, and should rank as an Imperfect Induction.

The same remarks will apply to those cases in which there is any uncertainty as to the nature of the fact of causation. If the inference be, say, that the two phenomena either are one cause and the other effect, or stand to each other in the relation of cause and effect, though we may be unable to determine which of the two is cause and which is effect, or are both of them effects of the same cause (adding any other alternatives which the particular case may require), the inference is, so far as it goes, a Perfect Induction. But, if one or some only of these alternatives be selected, on any grounds short of absolute or moral certainty, to the exclusion of the others, the inference is only probable, and must be regarded as merely an Imperfect Induction.

Briefly to sum up the contents of this chapter, Imperfect Inductions are the results either of an Inductio per Enumerationem Simplicem (to which I propose to appropriate the expression 'Empirical Generalisations'), or of the Argument from Analogy (which I call Analogies), or of an imperfect fulfilment of one or other of the Inductive Methods (to which we might, perhaps, advantageously appropriate the expression 'Incomplete Inductions'). In the two former cases there can be no more than an intimation of a Fact of Causation, while in the last we conceive ourselves to be on the way towards establishing one.

CHAPTER V.

On the relation of Induction to Deduction, and on Verification.

THE results of our inductions are summed up in general propositions, which are not unfrequently stated in the shape of mathematical formulæ. These general propositions, the results of inductive reasoning, become, in turn, the data from which deductive reasoning proceeds. Though the major premiss of any single deductive argument may itself be the result of deduction, it will invariably be found, as pointed out long ago by Aristotle[1], that the ultimate major premiss of a chain of deductive reasoning is a result of induction. There must be some limit to the generality of the propositions under which our deductive inferences can be subsumed, and, when we have reached this limit, the only evidence on which the ultimate major premiss can repose, if it depend on evidence at all, must be inductive. Thus, most of the deductions in the science of Astronomy, and

---

[1] ʽΗ μὲν δὴ ἐπαγωγὴ ἀρχή ἐστι καὶ τοῦ καθόλου, ὁ δὲ συλλογισμὸς ἐκ τῶν καθόλου. Εἰσὶν ἄρα ἀρχαὶ ἐξ ὧν ὁ συλλογισμός, ὧν οὐκ ἔστι συλλογισμός· ἐπαγωγὴ ἄρα.—*Eth. Nic.* vi. 3 (3). Cp. *Eth. Nic.* vi. 6 8 (9); *Metaphysics*, i. 1; *Posterior Analytics*, ii. 19.

R

many of those in the science of Mechanics, depend ultimately on the Law of Universal Gravitation; but this Law itself is the result of an induction based upon a variety of facts, including both the fall of bodies to the earth and the motion of the planets in their orbits. Again, a large number of geometrical deductions may be traced up to the ultimate major premiss: 'Things that are equal to the same thing are equal to one another.' But this proposition, if not referred directly to induction, is classed under the head of intuitive conceptions, the most probable, though perhaps not the most commonly received, explanation of which is that which derives them from the accumulated experience of generations, transmitted hereditarily from father to son.

A Deductive Inference combines the results of previous inductions or deductions, and evolves new propositions as the consequence, or, to put the matter in a slightly different point of view, as expressing the total result, of these combinations. I append a few easy examples of the manner in which the results of induction are employed in a deductive argument.

To begin with a very simple instance, but one which will serve as a good illustration of the stage at which our investigations cease to be inductive and become deductive;—suppose we, have ascertained, by previous inductions, that A produces a, B produces β, C produces $-\frac{a}{2}$, D produces $\frac{a}{2}$, and E produces $\frac{a}{3}$, we know, by calculation—that is, by deductive reasoning—that the total effect of A, B, C, D, E is $\beta + \frac{4a}{3}$. In this case the simple rules of Algebra, governing the addition and subtraction of quantities, combined with the special data here furnished, 'are the premises from which our deductive reasoning proceeds.

The proposition proved in Euclid, Book i. Prop. 38, that 'Triangles upon equal bases, and between the same parallels, are equal to one another,' is derived from, or is the total result of, the previous deductions (1) that 'Parallelograms upon equal bases, and between the same parallels, are equal to one another,' (2) that 'Triangles formed by the diagonal of a parallelogram are each of them equal to half the parallelogram' (i. 34), and (3) the previous induction that 'the halves of equal things are equal.'

What is called the Hydrostatic Paradox, namely, that a man standing on the upper of two boards, which form the ends of an air-tight leather bag, and blowing through a small tube opening into the space between the board, can easily raise his own weight, is a combination of two propositions, both gained from experience by means of induction, these propositions being (1) that fluids transmit pressure equally in all directions, (2) that, the greater the pressure brought to bear on any surface from below, the greater the weight which it will sustain (otherwise expressed by the Mechanical Law that action and reaction are equal).

To take another very simple instance of a similar kind. One of the earliest and easiest problems in the Science of Optics is the following: 'A conical pencil of rays is incident upon a plane reflecting surface; to determine the form of the reflected pencil.' The solution, that the reflected pencil will be a cone having for its vertex a certain imaginary point, which can be geometrically determined, on the other side of the surface, is derived from a combination of the experimental truth, gained by induction, that 'the angle of reflexion is equal to the angle of incidence' with the geometrical propositions stated in Euclid i. 8 and i. 29.

In the Science of Political Economy, Ricardo's Theory of Rent, when stated in the slightly modified form that 'the rent of land represents the pecuniary value of the advantages which such land possesses over the least valuable land in cultivation,' is an easy deduction from two principles which are supplied by every one's experience, namely, (1) that land varies in value, and (2) that there is some land either so bad or so disadvantageously situated as to be not worth the cultivating [2].

Professor Cairnes' work on the *Slave Power* furnishes a remarkable example of the successful application of the deductive method to the determination of economical questions. The economical effects of slavery are thus traced. We learn from observation and induction that slave labour is subject to certain characteristic defects:

---

[2] The student will find an easy exposition of this Theory in Fawcett's *Manual of Political Economy*, Bk. II. ch. iii. *ad init.* As originally stated, Ricardo's theory neglected to take account of advantages of situation, such as proximity to a market, and regarded the value of land as depending solely on its fertility.

it is given reluctantly; it is unskilful; and, lastly, it is wanting in versatility. As a consequence of these characteristics, it can only be employed with profit when it is possible to organise it on a large scale. It requires constant supervision, and this for small numbers or for dispersed workmen would be too costly to be remunerative. The slaves must, consequently, be worked in large gangs. Now there are only four products which repay this mode of cultivation, namely, cotton, sugar, tobacco, and rice. Hence a country in which slave labour prevails is practically restricted to these four products, for it is another characteristic of slave labour, under its modern form, that free labour cannot exist side by side with it. But, besides restricting cultivation to these four products, some or all of which have a peculiar tendency to exhaust the soil, slave labour, from its want of versatility, imposes a still further restriction. 'The difficulty of teaching the slave anything is so great —the result of the compulsory ignorance in which he is kept, combined with want of intelligent interest in his work — that the only chance of rendering his labour profitable is, when he has once learned a lesson, to keep him to that lesson for life. Accordingly, where agricultural operations are carried on by slaves, the business of each gang is always restricted to the raising of a single product. Whatever crop be best suited to the character of the soil and the nature of slave industry, whether cotton, tobacco, sugar, or rice, that crop is cultivated, and that crop only. Rotation of crops is thus precluded by the conditions of the case. The soil is tasked again and again to yield the same product, and the inevitable result follows. After a short series of years its fertility is completely exhausted, the planter abandons the ground which he has rendered worthless, and passes on to seek in new soils for that fertility under which alone the agencies at his disposal can be profitably employed.' Thus, from the characteristics of slave labour may be deduced the economical effect of exhaustion of the soil on which it prevails, and the consequent necessity of constantly seeking to extend the area of cultivation. From the peculiar character of the crops which can alone be successfully raised by slave labour may be explained the former prevalence of slavery in the Southern, and its absence in the Northern, States of the American Union; and from the necessity of constantly seeking fertile virgin soil for the employment of slave labour may be explained the former policy of the Southern States, which was invariably endeavouring to bring newly constituted States under the dominion of slave institutions [3].

These examples of the combination of inductive with deductive reasoning might be multiplied to any extent. Mechanics, Astronomy, and the Mathematico-physical sciences generally, furnish, perhaps, the most striking instances of it. The great importance of deduction as an instrument for the ascertainment of physical truths

[3] See Professor Cairnes on the *Slave Power*, ch. ii. His arguments are stated in a condensed form in Fawcett's *Manual of Political Economy*, Bk. II. ch. xi.

could hardly be illustrated more appropriately than by the following cases adduced by Sir John Herschel[4]:—

' It had been objected to the doctrine of Copernicus, that, were it true, Venus [and, it might have been added, Mercury, as the other inferior planet] should appear sometimes horned like the moon. To this he answered by admitting the conclusion, and averring that, should we ever be able to see its actual shape, it *would* appear so. It is easy to imagine with what force the application would strike every mind when the telescope confirmed this prediction, and showed the planet just as both the philosopher and his objectors had agreed it ought to appear. The history of science affords perhaps only one instance analogous to this. When Dr. Hutton expounded his theory of the consolidation of rocks by the application of heat, at a great depth below the bed of the ocean, and especially of that of marble by actual fusion; it was objected that, whatever might be the case with others, with calcareous or marble rocks, at least, it was impossible to grant such a cause of consolidation, since heat decomposes their substance and converts it into quicklime, by driving off the carbonic acid, and leaving a substance perfectly infusible, and incapable even of agglutination by heat. To this he replied, that the pressure under which the heat was applied would prevent the escape of the carbonic acid; and that being retained, it might be expected to give that fusibility to the compound which the simple quicklime wanted. The next generation saw this anticipation converted into an observed fact, and verified by the direct experiments of Sir James Hall, who actually succeeded in melting marble, by retaining its carbonic acid under violent pressure.'

It should be noticed that, for the most part, in the actual conduct of scientific enquiry, there is a constant

[4] *Discourse on the Study of Natural Philosophy*, § 299.

alternation of the processes of Induction and Deduction. A truth obtained inductively is often at once used, either by itself or in combination with other propositions, for the purpose of evolving new truths by deduction, while it may also be subsequently employed together with other inductions of the same order for the purpose of leading up inductively to propositions of a higher degree of generality. We are constantly passing from the one process to the other, and back again, and often it becomes exceedingly difficult to determine exactly how much of our ultimate conclusion is due to the one method, and how much to the other. It is an error (though this error has received the countenance of Bacon) to suppose that the process of induction should always be pursued continuously up to a certain point, and that from that point the process of deduction should proceed equally uninterruptedly. We may, and in fact should, frequently pause to consider to what deductive conclusions our inductive inferences lead, or to try whether they may not be connected by a chain of deductive reasoning with wider truths previously ascertained[5].

A very common instance of the constant interlacing of the inductive and deductive processes just noticed is to be found in the ordinary mode of framing and employing hypotheses. First, our hypotheses are always suggested by some fact, or facts, within our experience. They are thus based on a rough kind of induction. When framed,

---

[5] On this subject, see the excellent criticism on Bacon in Mr. Mill's *Logic*, Bk. VI. ch. v. § 5.

we generally proceed to trace the consequences which would ensue on the supposition of their truth. This is a deductive process. Individual facts or inductions from individual facts are then compared with these results, and, if they agree with them, are regarded as confirmatory of the hypothesis. Of course, these processes may be frequently repeated, and are often so repeated, the hypothesis thus constantly gaining in probability, even though it may as yet have no claim to be regarded as an established truth. Lastly, if it attain the position of a valid induction, it must be by the application of one or other of the inductive methods, which converts its previous probability into scientific certainty. Or, perhaps, it may be finally established not by induction at all, but by being brought deductively under some more general law.

These remarks and the instances adduced above naturally lead to a discussion of the place to be assigned in scientific enquiry to the process called Verification. In Deductive Reasoning, especially when it involves elaborate calculations, there is always great danger lest we should have omitted to take into account some particular agency or element, or have miscalculated its effects, or have formed a false estimate of the combined effect of the various agencies or elements in operation. The only remedy against these possible errors, besides the employment of great caution in the conduct of the deductive process itself, is to be found in Verification, a word which, in its stricter sense, appears to be applied to the process of testing, by means of an appeal to facts, the validity of the conclusions already arrived at by a course of deductive reasoning. Thus it had been deductively inferred from the Copernican theory that the planets Venus and Mercury ought -to pass through phases, like the moon, and the application of the telescope, by means of which they were actually seen to assume these phases, furnished a triumphant verification of the inference. Every occurrence of an eclipse of the sun or moon or of the transit or occultation of a star, when it accords with the previous calculations of astronomers, is also an instance of Verification in this, the stricter, sense of the term. The discovery of the planet Neptune affords an excellent instance of the same kind. But the word is often used in a looser sense and extended to all cases in which an appeal is made to facts, as, for instance, when we perform an experiment in order to test the truth of a hypothesis, or where we employ the Method of Difference in order to supplement the characteristic uncertainty attaching to the employment of the Method of Agreement. Of the process denoted by this looser sense of the word, instances will readily occur to every one. Thus, the diminution in the periods of Encke's comet has been regarded by some astronomers (though, perhaps, erroneously) as a verification of the theory that space is filled with an interstellar medium; or, to take an instance from a very different class of subjects, the recent breaking-up of the slave-system in the Southern States of

America may be regarded as a verification of the prediction that slave and free institutions could not long co-exist under the same political form of government. For an instance of a case in which the Method of Difference is called in to verify a previous employment of the Method of Agreement, I may refer back to the' enquiry into the cause of crystallization, already adduced in my discussion of those two methods[6].

There is a still wider application of the word Verification, by which it is extended to any corroboration of one mode of proof by means of another. It thus includes a deductive proof adduced in corroboration of an inductive one. The most common instance of this kind of verification is the inclusion of a partial under a more general law, the partial law having been arrived at inductively, and it being subsequently shown that the more general law leads deductively to it. Thus, the phenomena of the Tides had, prior to the epoch of Newton, been partially explained by the inductive method. Newton, by deducing these phenomena from the Law of Universal Gravitation, not only afforded a much more complete explanation, but also furnished the most convincing verification of the results already arrived at. Similarly, the laws of falling bodies on the earth's surface, which had already been proved inductively, were, from the time of Newton, brought under the law of universal gravitation, and proved deductively from it. The same was also the case with Kepler's Laws, when they were proved

[6] See pp. 145, 146, 157, 158.

deductively from the theorem of the Central Force. This mode of Verification is recommended by Mr. Mill, under the name of the Inverse Deductive or Historical Method, as specially applicable to generalisations on society which have been inferred inductively from the study of history ‘or the observation of mankind. These generalisations are subsequently verified by being connected deductively with the general laws of mind or conduct which are furnished by the study of Psychology or Ethology[7]. It is thus shown that the generalisations of history are such as we might have anticipated *à priori* from a general knowledge of human nature, and each branch of the enquiry is made in this manner to afford a striking confirmation of the results arrived at by the other.

It frequently happens that what may be called a residual phenomenon affords an unexpected, and, on that account, a striking verification of some law which is not immediately the object of investigation. Thus, to recur to an instance already adduced for another purpose, when it was found that the difference between the observed and calculated velocities of sound was exactly

[7] See above, pp. 204–207, and Mill's *Logic*, Bk. VI. ch. x. I cannot agree with Mr. Mill in attaching any special importance to the *order* in which the respective Methods are used in this enquiry. Though the inductive investigation, based on the facts of history or observation, generally precedes the deductive verification from the laws of psychology, we may, and sometimes do, begin with psychological generalisations, and subsequently verify them by an appeal to observed facts. The only essential point is, that the two Methods should be combined, so that the results arrived at by the one may corroborate the results arrived at by the other.

accounted for by the law of the development of heat by compression, this law acquired so novel and striking a confirmation as to leave no doubt of its truth or universality.

It need hardly be remarked that any verification of one inductive proof by another, or of an induction by a deduction, or of a deduction by an induction, should conform with the laws of deductive or inductive reasoning as the case may be. Verification is not a distinct mode of proof, but is simply a confirmation of one proof by another, sometimes of a deduction by an induction, sometimes of an induction by a deduction, and, finally, sometimes of one induction or deduction by another. It must also be borne in mind that the term is not infrequently employed to designate simply the confirmation of a hypothesis by an appeal to facts.

The student will, of course, understand that it is not always necessary to employ Verification. A proof may be so cogent as to need no confirmation. It would be absurd, for instance, to appeal to actual measurement as a verification of the proposition enunciated in *Euclid*, i. 47.

CHAPTER VI.

On the Fallacies incident to Induction.

THE errors incidental to inductive reasoning and to its various subsidiary processes have already, to a great extent, been noticed in the preceding chapters. In laying down the conditions essential to the correct conduct of a process, the mistakes which result from its incorrect conduct necessarily form part of our enquiry. Though, therefore, it may be convenient to pass the inductive fallacies in review, it is assumed that the student is already acquainted with the principal errors to which his processes and methods are liable.

A. To begin with the subsidiary processes, the errors incident to the process of observation, or 'the fallacies of mis-observation,' are well classified by Mr. Mill as those which arise from Non-observation and those which arise from Mal-observation.

I. Non-observation may consist either (1) in neglecting some of the instances, or (2) in neglecting some of the circumstances attendant on a given instance.

(1) With respect to the non-observation of instances, it

was long ago pointed out by Bacon [1] that there is in the human mind a peculiar tendency to dwell on affirmative and to overlook negative instances. Familiar examples of this tendency will readily occur to every one. We think it a ' curious coincidence ' that we should suddenly meet a man of whom we have just been talking, that some event should happen of which we dreamed the night before, or that the predictions of a fortune-teller or an almanac should be verified by the facts. The explanation of these ' curious coincidences ' is that our attention is arrested by the affirmative instances, whereas

[1] ' Intellectus humanus in iis quæ semel placuerunt (aut quia recepta sunt et credita, aut quia delectant) alia etiam omnia trahit ad suffragationem et consensum cum illis : et licet major sit instantiarum vis et copia, quæ occurrunt in contrarium ; tamen eas aut non observat, aut contemnit, aut distinguendo summovet et rejicit, non sine magno et pernicioso prejudicio, quo prioribus illis syllepsibus auctoritas maneat inviolata. Itaque recte respondit ille, qui, cum suspensa tabula in templo ei monstraretur eorum qui vota solverant quod naufragii periculo elapsi sint, atque interrogando premeretur, anne tum quidem Deorum numen agnosceret, quæsivit denuo, "At ubi sunt illi depicti qui post vóta nuncupata perierint?" Eadem ratio est fere omnis superstitionis, ut in astrologicis, in somniis, ominibus, nemesibus, et hujusmodi ; in quibus homines delectati hujusmodi vanitatibus advertunt eventus, ubi implentur ; ast ubi fallunt, licet multo frequentius, tamen negligunt et prætereunt. At longe subtilius serpit hoc malum in philosophiis et scientiis ; in quibus quod semel placuit reliqua (licet multo firmiora et potiora) inficit, et in ordinem redigit. Quinetiam licet abfuerit ea, quam diximus, delectatio et vanitas, is tamen humano intellectui error est proprius et perpetuus, ut magis moveatur et excitetur affirmativis, quam negativis ; cum rite et ordine æquum se utrique præbere debeat ; quin contra, in omni axiomate vero constituendo, major est vis instantiæ negativæ.' —*Novum Organum*, Lib. I. Aph. xlvi.

the numberless instances in which there is no correspondence between the one set of facts and the other altogether escape our notice. We probably talk scores of times during the day of persons whom we do not meet immediately afterwards; we frequently dream in the most circumstantial manner of events which never occur; and, where one prediction of a fortune-teller is verified, scores are probably falsified. The weather-prophets of the almanacs possess a considerable advantage in the fact that, whereas, at all times, there is at least a considerable chance of their predictions turning out true, there are certain periods, such as the equinoxes, at which particular kinds of weather may be anticipated with a probability amounting almost to certainty.

In former generations 'coincidences' of this kind were regarded not simply as 'curious' and 'remarkable,' but as proofs of some causal connexion between the events. To talk of a person was supposed to render his presence more likely; a verified prediction was regarded as evidence of second-sight; and a comet which was observed to be followed by a war was supposed to be, if not the cause of the war, at least a messenger sent from Heaven to proclaim its approach. The tendency to take note of affirmative, and to overlook negative instances, is one of the causes of that hasty generalisation of which I shall speak in a subsequent part of this chapter[2].

[2] The following remarks of Sir John Herschel, in speaking of the verification of 'signs of the weather,' are so apposite, that I append them in a note.

This tendency is considerably intensified, if the affirmative instances are regarded as illustrations of some preconceived theory [3], or if the evidence afforded by them be supplemented by some powerful affection of the mind [4]. It seldom happens that men can hold themselves entirely indifferent with respect to two rival

'We would strongly recommend any of our readers whose occupations lead them to attend to the "signs of the weather," and who, from hearing a particular weather adage often repeated, and from noticing themselves a few remarkable instances of its verification, have "begun to put faith in it," to commence keeping a note-book, and to set down without bias all the instances which occur to them of he recognised antecedent, and the occurrence or non-occurrence of the expected consequent, not omitting also to set down the cases in which it is left undecided; and, after so collecting a considerable number of instances (not less than a hundred), proceed to form his judgment on a fair comparison of the favourable, the unfavourable, and the undecided cases; remembering always that the *absence of a majority one way or the other would be in itself an improbability*, and that, therefore, to have any weight, the majority should be a very decided one, and *that* not only in itself, but in reference to the neutral instances. We are all involuntarily much more strongly impressed by the fulfilment than by the failure of a prediction, and it is only, when thus placing ourselves face to face with fact and experience, that we can fully divest ourselves of this bias.'—*Familiar Lectures on Scientific Subjects*, Lecture IV.

[3] 'Habet enim unusquisque (præter aberrationes naturæ humanæ in genere) specum sive cavernam quandam individuam, quæ lumen naturæ frangit et corrumpit; . . . . vel propter differentias impressionum, prout occurrunt in animo præoccupato et prædisposito, aut in animo æquo et sedato.'—Bacon's *Novum Organum*, Lib. I. Aph. xlii.

[4] 'Intellectus humanus luminis sicci non est; sed recipit infusionem a voluntate et affectibus; id quod generat *ad quod vult scientias:* quod enim mavult homo verum esse, id potius credit.'—*Novum Organum*, Lib. I. Aph. xlix.

S

opinions and apply themselves to the comparatively
unexciting task of collecting evidence impartially. on
either side.   To avoid taking a side on imperfect information, even where our interests or passions are not directly
concerned, is one of the last and most difficult lessons
learned by the scientific intellect, and by ordinary men
it is regarded as a sign of a peculiarly frigid temperament, if not of an indifference to truth.   Thus, when
the theory involved in the idea of witchcraft had once
been conceived and accepted, and especially when it
had led to the invention of a new crime, it came to
be held that the burden of proof lay with those who
called its reality in question.   Every story which confirmed the theory would be greedily received, while
instances in which the supposed powers of the witch
had failed, if noticed at all, would either leave but
a slight impression on the mind, or be easily accounted for by supposing the intervention of a higher
power.   To the numerous class engaged in the administration of the laws, a not unnatural reluctance
to question the justice of the principles on which they
and their predecessors had been in the habit of acting would furnish an additional inducement to pass
lightly over negative instances.   Fear, or dread of
eccentricity, would operate in the case of others;
and thus a theory of the most preposterous character,
which, to a mind not preoccupied, received little ·or no
confirmation from facts[5], and the truth of which could

---

[5] When a person was convinced that he was subject to the evil

easily have been brought to the test, maintained its ground, and throughout many centuries continued to produce the most mischievous results. The extent of the bias to which the mind, in its observation of instances, is exposed from the influence of strong affections, is patent to every one. A man of a desponding temperament will dwell on the number of those who have failed, a man of a sanguine temperament on the number of those who have succeeded, in their respective professions. A man with strong sympathies will see only virtues or good traits of character, where one of a malevolent or critical disposition will see only vices or blemishes. An ardent adherent of a religious sect or a political party will see nothing but good in those who agree with him, nothing but evil in those who adopt a different creed or profess to be guided by different principles of policy.

Many of the above errors might be otherwise described as arising from the confusion between *absolute* and *relative frequency*. We notice how often an event occurs, but we do not notice how much oftener it does not occur.

Not only will a preconceived opinion or a powerful affection come in aid of the natural tendency of men to dwell on affirmative and overlook negative instances, but they will often cause them to adhere to theories for which, whatever may have been the history of their formation,

practices of a witch, this conviction would, of course, sometimes produce the ill effects attributed to witchcraft itself. In other cases, some event, such as a death or an illness, which occurred in the ordinary course of nature, would confirm the suspicion.

there is absolutely no support whatever in fact. Thus, the theory which prevailed down to the time of Galileo[6], that bodies fall to the earth in times inversely proportional to their weights, so that a body weighing, say, five pounds, would fall in a time five times as short as a body weighing one pound, rested on absolutely no evidence except the fact that, in consequence of the resistance of the air, the heavier body, especially if it be of a denser material, reaches the ground in a shorter time than the lighter one; still, till Galileo made his experiments, at the end of the sixteenth century, from the leaning tower of Pisa, no one thought of bringing to a decisive test a theory which it was so easy to prove or disprove. Even, without having recourse to experiment, one would have imagined that the most casual observations of falling bodies would have revealed, to a mind not strongly pre-occupied, the strange inaccuracy of this theory. The reception accorded to the theory that the weight of the elements increases in a tenfold ratio, so that earth is ten times heavier than water, water ten times heavier than air, and air ten times heavier than fire, seems still more astounding[7].

In Sir Thomas Browne's *Enquiries into Vulgar and Common Errors*[8], we have an examination of the propo-

---

[6] *Galilæi Systema Cosmicum*, Dial. II.

[7] This theory appears to have originated in a mistaken interpretation of a passage in Aristotle, *De Generatione et Corruptione*, II. 6. See my note on Bacon's *Novum Organum*, Lib. I. Aph, xlv.

[8] Bk. IV. ch. vii.

sition that 'men weigh heavier dead than alive, and before meat than after.' Here are two extraordinary paradoxes which it was perfectly easy for any one to bring to a decisive test; and still, though an appeal to facts would at once have been fatal to them, they appear to have met with a very general reception. The grounds assigned for the prevalence of the latter opinion are so curious that they deserve to be transcribed. 'Many are also of opinion, and some learned men maintain, that men are lighter after meals than before, and that by a supply and addition of spirits obscuring the gross ponderosity of the aliment ingested; but the contrary hereof we have found in the trial of sundry persons in different sex and ages. And we conceive men may mistake, if they distinguish not the sense of levity unto themselves, and in regard of the scale, or decision of trutination. For after a draught of wine, a man may seem lighter in himself from sudden refection, although he be heavier in the balance, from a corporal and ponderous addition; but a man in the morning is lighter in the scale, because in sleep some pounds have perspired; and is also lighter unto himself, because he is refected.' It will be noticed that 'spirits' are supposed to possess the property of positive levity, and that, consequently, they are regarded as making any body into which they enter lighter than it was before. The theory that certain bodies are positively light is itself an instance of a fallacy of non-observation, but, as will be seen presently, of non-observation of circumstances not of instances.

Another extraordinary instance of a statement which obtained acceptance without any foundation whatever in fact is noticed in an article in the *Quarterly Review* for January, 1865, on 'Aristotle's History of Animals.' Here, however, there appears to be no assignable reason for the mistake.

'Aristotle held some peculiar notions with respect to the skull. He says, " that part of the head which is covered with hair is called the cranium ; the fore part of this is called the sinciput ; this is the last formed, being the last part in the body which becomes hard." He correctly alludes here to the opening in the frontal bone of a young infant, which gradually becomes hardened by ossification ; "the hinder part is the occiput, and between the occiput and sinciput is the crown of the head : the brain is placed beneath the sinciput, and the occiput is empty (!). The skull has sutures : in women there is but one placed in a circle (!) ; men have generally three joined in one, and a man's skull has been seen without any sutures at all." The often repeated question as to how far Aristotle's observations are the result of his own investigation, naturally suggests itself again here ; had Aristotle ever dissected a human body, he never would have asserted a proposition so manifestly false as that the back of the head is empty, or that women have one only suture placed in a circle.'

The passage here noticed occurs in the *Historia Animalium*, Bk. I. ch. vii. Cp. Bk. III. ch. vii.

A still more remarkable instance of this description of fallacy is noticed in Mr. Lecky's *History of European Morals from Augustus to Charlemagne*[9].

[9] Vol. i. p. 394.

'Aristotle, the greatest naturalist of Greece, had observed that it was a curious fact, that on the sea-shore no animal ever dies except during the ebbing of the tide. Several centuries later, Pliny, the greatest naturalist of an empire that was washed by many tidal seas, directed his attention to this statement. He declared that, after careful observations which had been made in Gaul, it had been found to be inaccurate, for what Aristotle stated of all animals, was in fact only true of man. It was in 1727 and the two following years, that scientific observations made at Rochefort and at Brest finally dissipated the delusion.'

I add one more instance, showing the extraordinary readiness with which men, even of remarkable acuteness and erudition, will accept the strangest fancies, though absolutely unsupported by evidence. It is taken from Glanvill's *Scepsis Scientifica*, published in 1665 [10] :—

'Besides this there is another way of secret conveyance that's whisper'd about the World, the *truth* of which I vouch not, but the *possibility :* it is conference at distance by sympathized handes. For say the relatours of this strange secret: The hands of two friends being allyed by the transferring of Flesh from one into the other, and the place of the Letters mutually agreed on; the least prick in the hand of one, the other will be sensible of, and that in the same part of his own. And thus the distant friend, by a new kind of Chiromancy, may read in his own hand what his correspondent had set down in his. For instance, would I in London acquaint my intimate in Paris, that I am well: I would then prick that part where I had appointed the letter'[I] and, doing so in another place to signifie that word was done, proceed to [A],

---

[10] *Scepsis Scientifica*, ch. 24.

thence to [M], and so on, till I had finisht what I intended to make known.'

The influence of some strong passion or affection in causing men to accept theories without any support from observation or experiment, and often in direct defiance of them, may be illustrated from almost all the more powerful feelings of human nature. The mythologies of every nation are full of the wildest and most improbable stories, originating partly in the strength of the religious sentiment, partly in that love of the marvellous which seems to be connatural to every race of mankind, partly in later misinterpretations of that poetical language in which early races are wont to clothe their ideas. Thus, stories of the transformation of men into beasts, of rivers flowing backwards, of images falling down from heaven, besides other tales still more fantastic, have been greedily received by generation after generation, in spite of all the analogies of nature and without one single instance to confirm them. The beliefs in ghosts, spirit-rapping, and similar phenomena, seem to have their origin in man's insatiable craving for the marvellous, acting often in combination with the feelings of fear, hope, or curiosity. One of the most powerful agents in human affairs is the passion of avarice or the insatiable desire for the accumulation of wealth. In the middle ages, this passion led the alchemists, contrary to all experience, to the belief that it was open to men to become suddenly and enormously rich by discovering the secret of transmuting other metals into gold. In modern times it has frequently led, and still leads, men to embark in the most desperate speculations, which no scientific calculation of chances would justify. In a lottery, for instance (which is a comparatively innocuous form of speculation), the value of the chance is, owing to the expenses of management and the profit required by the projectors, invariably much less than the price paid for the ticket. But, perhaps, the most remarkable exemplification of the unreasoning desire for sudden accessions of wealth is to be found in the pertinacity with which, in spite of every warning, men would, till within the last few years, expend large fortunes in sinking shafts for coal and other minerals in strata in which the universal experience of geologists and miners testified against their occurrence. In this, as in many other cases, the observations of competent authorities went for nothing; the passion was so absorbing that it alone determined action.

The fallacies due to non-observation of instances may be further exemplified by the tendency of the mind to acquiesce in the first instances which offer themselves [11],  especially if they be of a striking kind [12], instead of care-

---

[11] 'Axiomata, quæ in usu sunt, ex tenui et manipulari experientia, et paucis particularibus, quæ ut plurimum occurrunt, fluxere ; et sunt fere ad mensuram eorum facta et extensa.'—*Novum Organum,* Lib. I. Aph. xxv.

[12] 'Intellectus humanus illis, quæ simul et subito mentem ferire et subire possunt, maxime movetur ; a quibus phantasia impleri et inflari consuevit : reliqua vero modo quodam, licet imperceptibili, ita se habere fingit et supponit, quomodo se habent pauca illa quibus mens

fully searching for other instances of a similar nature with which to compare and by means of which to interpret them. Thus, the phenomena of thunder and lightning would probably have received a much earlier explanation, had the attention of men been sooner directed to the instances of electricity which nature presents of a less striking kind and on a smaller scale. Again, the difficulties presented to early speculators by volcanoes and earthquakes would have been considerably diminished, had they been aware of the fact that there is hardly any portion of the earth's surface which is not undergoing a constant change of level by the process either of elevation or of subsidence, though such change is usually imperceptible to each single generation. The mistakes originating in this source of error are countless. We observe certain peculiarities in some particular representative of a class, profession, or nation, and then proceed to argue as if all the members of the class, profession, or nation were like him. Or, a person on his travels in some country is unfavourably impressed with the hotel-keepers, porters, and carriage-drivers, and then proceeds to denounce the whole nation to which they belong, as if the characteristics of a few exceptional classes were the characteristics of a nationality [13].

obsidetur; ad illum vero transcursum ad instantias remotas et heterogeneas, per quas axiomata tanquam igne probantur, tardus omnino intellectus est et inhabilis, nisi hoc illi per duras leges et violentum imperium imponatur.'—*Novum Organum*, Lib. I. Aph. xlvii.

[13] The history of the French language furnishes a striking instance of non-observation and of the curious and baseless theories to which

The student must have already perceived that I am trenching on Fallacies of Generalisation. When we proceed to treat insufficient evidence, or the absence of evidence, or popular beliefs which run counter to all

it may lead : ' It is well known that before certain feminine substantives, such as *messe, mère, soif, faim, peur,* &c., the adjective *grand* keeps its masculine termination, *grand'messe, grand'mère,* &c. Why so ? Grammarians, who are puzzled by nothing, tell us without hesitation that *grand* is here put for *grande,* and that the apostrophe marks the suppression of the final *e.* But the good sense of every scholar protests against this : after having learnt in childhood that *e* mute is cut off before a vowel, and never before a consonant, he is told that the *e* is here cut off without the slightest reason in such phrases as *grand'route,* &c. The real explanation is in fact a very different one. In its beginning, French grammar was simply the continuation and prolongation of Latin grammar ; consequently the Old French adjectives followed in all points the Latin adjective ; those adjectives which had two terminations for masculine and feminine in Latin (as *bonus, bona*) had two in Old French, whereas those Latin adjectives which had but one (as *grandis, fortis,* &c.) had only one in French. In the thirteenth century men said *une grand femme,* grandis femina ; *une âme mortel,* anima mortalis ; *une coutume cruel,* consuetudo crudelis ; *une plaine vert,* planities viridis, &c. In the fourteenth century the meaning of this distinction was no longer understood ; and men, deeming it a mere irregularity, altered the form of the second to that of the first class of adjectives, and wrote *grande, verte, forte,* &c., after the pattern of *bonne,* &c. A trace of the older and more correct form survives in such expressions as *grand'mère, grand'route, grand'faim, grand'garde,* &c., which are the *débris* of the older language. In the seventeenth century, Vaugelas and the grammarians of the age, in their ignorance of the historic reason of this usage, pompously decreed that the form of these words arose from an *euphonic* suppression of the *e* mute, which must be indicated by an apostrophe.'—Brachet's *Historical Grammar of the French Tongue,* Dr. Kitchin's Transl., Preface, p. vi ; Seventh Edition, pp. xxvi-vii

the evidence available, as if they afforded perfectly sufficient evidence, the fallacy is one of inference, and, if
the simulated inference be inductive, it is a Fallacy of
Generalisation. But the absence or insufficiency of the
evidence, if due to our not having kept our minds
sufficiently open to facts or not having taken sufficient
pains to collect all the facts pertinent to the question,
is a Fallacy of Non-observation, and is a defect in the
preliminary process rather than in the inductive inference itself. All the instances described above, I believe,
fall under this head, though the inferences founded upon
them, where they possess any show of justification at all,
are cases of unwarranted Inductio per Enumerationem
Simplicem, and so afford illustrations of Fallacies of
Generalisation.

(2) The second division of the Fallacies of Non-observation is the fallacy which arises from overlooking some
of the material circumstances attendant on a given instance. Here the defect is not in the number or per-
tinency of the instances, but in their character; the
description of the instances themselves is untrustworthy.
Till we have ascertained that we are fully acquainted with
all the material circumstances of the cases examined, we
cannot rely upon our facts, and, consequently, we have
no right to proceed to ground any inference upon them.
'The circumstances,' says Sir John Herschel[14], 'which

[14] Herschel's *Discourse on the Study of Natural Philosophy*,
§ 111.

accompany any observed fact, are main features in its
observation, at least until it is ascertained by sufficient
experience what circumstances have nothing to do with
it, and might therefore have been left unobserved without
sacrificing *the fact*. In observing and recording a fact,
therefore, altogether new, we ought not to omit any cir-
cumstance capable of being noted, lest some one of the
omitted circumstances should be essentially connected
with the fact. . . . For instance, in the fall of meteoric
stones, flashes of fire are seen proceeding from a cloud,
and a loud rattling noise like thunder is heard. These
circumstances, and the sudden stroke and destruction
ensuing, long caused them to be confounded with an
effect of lightning, and called thunderbolts. But one
circumstance is enough to mark the difference : the flash
and sound have been perceived occasionally to emanate
from a *very small cloud* insulated in *a clear sky ;* a combination of circumstances which never happens in a
thunder storm, but which is undoubtedly intimately connected with their real origin.'

The extreme difficulty of obtaining, by means of the
thermometer, a correct measure of the temperature of
the atmosphere, owing to the conduction of heat by the
stand and its radiation from surrounding objects, and the
consequent errors frequently made by observers from not
sufficiently providing against, or allowing for, these
sources of interference, will serve to every one as a
familiar illustration of the great importance of the caution
which it is here intended to furnish.

The following examples, adduced by Mr. Mill[15], are so interesting and appropriate, that I take the liberty of transcribing them :—

'Such, for instance [namely, the imperfect observation of particular facts], was one of the mistakes committed in the celebrated phlogistic theory ; a doctrine which accounted for combustion by the extrication of a substance called phlogiston, supposed to be contained in all combustible matter. The hypothesis accorded tolerably well with superficial appearances : the ascent of flame naturally suggests the escape of a substance ; and the visible residuum of ashes, in bulk and weight, generally falls extremely short of the combustible material. The error was non-observation of an important portion of the actual residue, namely, the gaseous products of combustion. When these were at last noticed and brought into account, it appeared to be an universal law that all substances gain instead of losing weight by undergoing combustion ; and, after the usual attempt to accommodate the old theory to the new fact by means of an arbitrary hypothesis (that phlogiston had the quality of positive levity instead of gravity), chemists were conducted to the true explanation, namely, that, instead of a substance separated, there was on the contrary a substance absorbed.

'Many of the absurd practices which have been deemed to possess medicinal efficacy, have been indebted for their reputation to non-observance of some accompanying circumstance which was the real agent in the cures ascribed to them. Thus, of the sympathetic powder of Sir Kenelm Digby : "Whenever any wound had been inflicted, this powder was applied to the weapon that had inflicted it, which was, moreover, covered with ointment, and dressed two or three times a day. The wound itself, in the meantime, was directed to

---

[15] *System of Logic*, Bk. V. ch. iv. § 4.

be brought together, and carefully bound up with clean linen rags, but, *above all, to be let alone* for seven days, at the end of which period the bandages were removed, when the wound was generally found perfectly united. The triumph of the cure was decreed to the mysterious agency of the sympathetic powder which had been so assiduously applied to the weapon, whereas it is hardly necessary to observe that the promptness of the cure depended upon the total ex-clusion of air from the wound, and upon the sanative opera-tions of nature not having received any disturbance from the officious interference of art. The result, beyond all doubt, furnished the first hint which led surgeons to the improved practice of healing wounds by what is technically called the *first intention*[16]."'

The next example I extract from Bp. Wilkins' very curious tractate, entitled *A Discovery of a New World, or a Discourse tending to prove that 'tis probable there may be another Habitable World in the Moon:—*

' He [that is, 'a late reverend and learned Bishop,' writing ' under the feigned name of Domingo Gonsales'[17]] supposeth that there is a natural and usual passage for many creatures betwixt our earth and this planet. Thus, he says, those great multitude of locusts, wherewith divers countries have been destroyed, do proceed from thence. And if we peruse the authors who treat of them, we shall find that many times they fly in numberless troops or swarms, and for sundry days together before they fall are seen over those places in great high clouds, such as, coming nearer, are of extension enough

[16] Dr. Paris' *Pharmacologia*, pp. 23–24.

[17] The small tract here referred to is republished in vol. viii. of the *Harleian Miscellanies* (Park's Edition). The author was Francis Godwin, afterwards Bishop of Hereford, and author of the well-known book *De Præsulibus Angliæ Commentarius*.

to obscure the day, and hinder the light of the sun. .From which, together with divers other such relations, he concludes that 'tis not altogether improbable they should proceed from the moon. Thus, likewise, he supposes the swallows, cuckoos, nightingales, with divers other fowl, which are with us only half a year, to fly up thither when they go from us. Amongst which kind, there is a wild swan in the East Indies, which at certain seasons of the year do constantly take their flight thither. Now, this bird being of a great strength, able to continue for a long flight, as also going usually in flocks like our wild geese, he supposeth that many of them together might be thought to carry the weight of a man ; especially if an engine were so contrived (as he thinks it might) that each of them should bear an equal share in the burden. So that, by this means, 'tis easily conceivable how once a year a man might finish such a voyage ; going along with these birds at the beginning of winter, and again returning with them in the spring[18].'

A more accurate and extended series of observations would, of course, have shown that the birds and locusts migrated from other parts of the earth's surface.

It is not necessary to multiply examples of the errors arising from slovenliness and inattention in the collection or examination of our instances. The necessity of maintaining the strictest caution and accuracy in the conduct of our observations and experiments has already been insisted upon in the Second Chapter of this work.

II. Besides the errors which originate in the neglect of instances or of some of the circumstances which are con-

[18] Wilkins' *Discovery of a New World*, Fifth Edition, p. 160.

nected with a given instance, there is another class of errors derived from mistaking for observation that which is not observation at all, but inference. To this class of errors Mr. Mill gives the name of Fallacies of Mal-Observation. That which is strictly matter of perception does not admit of being called in question; it is the ultimate basis of all our reasoning, and, if we are to repose any confidence whatever in the exercise of our faculties, must be taken for granted. But there are few of our perceptions, even of those which to the unphilosophical observer appear to be the simplest, which are not inextricably blended with inference. Thus, as is well known to every student of psychology, in what are familiarly called the perceptions of distance and of form, the only perception proper is that of the various tints of colour acting on the retina of the eye, and it is by a combination of this with perceptions of touch, and of the muscular sense, that the mind gains its power of determining form and distance. Now, a judgment of this kind, which is really due to inference, is, especially by the uneducated and unreflecting, perpetually mistaken for that which is due to direct observation; and thus what is really only an inference from facts is often emphatically asserted to be itself a matter of fact. 'In proportion,' says Mr. Mill [19], 'to any person's deficiency of knowledge and mental cultivation, is generally his inability to discriminate between his inferences and the perceptions on which they were grounded. Many a marvellous tale,

[19] Mill's *Logic*, Bk. V. ch. iv. § 5.

T

many a scandalous anecdote, owes its origin to this incapacity. The narrator relates, not what he saw or heard,
but the impression which he derived from what he saw or
heard, and of which perhaps the greater part consisted
of inference, though the whole is related not as inference
but as matter-of-fact. The difficulty of inducing witnesses to restrain within any moderate limits the intermixture of their inferences with the narrative of their
perceptions, is well known to experienced cross-ex-
aminers; and still more is this the case when ignorant
persons attempt to describe any natural phenomenon.
" The simplest narrative," says Dugald Stewart, " of the
most illiterate observer involves more or less of hypothesis; nay, in general, it will be found that, in proportion to his ignorance, the greater is the number of
conjectural principles involved in his statements. A
village apothecary (and, if possible, in a still greater
degree, an experienced nurse) is seldom able to describe
the plainest case, without employing a phraseology of
which every word is a theory: whereas a simple and
genuine specification of the phenomena which mark a
particular disease; a specification unsophisticated by
fancy, or by preconceived opinions, may be regarded as
unequivocal evidence of a mind trained by long and
successful study to the most difficult of all arts, that of
the faithful *interpretation* of nature."

No better instance of the Fallacy of Mal-observation
can be given than what was called the common-sense
argument against the truth of the Copernican System.

That the earth should move round the sun, men said, was impossible; for, every day, they saw the sun rise and set and perform his course in the heavens. They felt the earth at rest, they saw the sun in motion, and it was absurd to call upon them to disbelieve the direct evidence of their senses. It need hardly be said that what they mistook for the direct evidence of their senses was really an inference. What they saw was consistent with one or other of two hypotheses, that the sun moved, or that the earth moved; and, neglecting to take any account of the latter, they assumed the former. If it were not for the impressions of a contrary kind derived from the actual motion of the carriage, a man, whirled along in a railway train, might with equal justice maintain, by an appeal to the evidence of his eyesight, that the trees and the houses were running past him.

Ventriloquism supplies another familiar instance of the same error. A man who had never before been imposed upon by the tricks of a ventriloquist, and who was not aware of the character of the deception, would be positive in maintaining that he had the direct evidence of the sense of hearing in support of his belief that the sound he heard proceeded from a particular person or a particular part of the building other than that from which it really came. The fact, of course, is that the sound itself is all that is directly perceived by the sense of hearing; the reference of it to a particular person or a particular place is an act of inference grounded upon constant, or at least frequent, association. What is done by the ventriloquist is not to deceive the sense of hearing, but to mislead the faculty of judgment.

What are called 'delusions' and 'hallucinations' furnish a further instance of Mal-observation. It seems to be now pretty generally agreed that these are due to morbid affections of the sensory ganglia. 'The patient's senses,' says Dr. Maudsley [20], speaking of what he calls *sensorial insanity*, ' are possessed with hallucinations, their ganglionic central cells being in a state of convulsive action; before the eyes are blood-red flames of fire, amidst which whosoever happens to present himself, appears as a devil, or otherwise horribly transformed; the ears are filled with a terrible roaring noise, or resound with a voice imperatively commanding him to save himself; the smell is perhaps one of sulphurous stifling; and the desperate and violent actions are, like the furious acts of the elephant, the convulsive reactions to such fearful hallucinations. The individual in such a state is a machine set in destructive motion, and he perpetrates the extremest violence or the most desperate murder without consciousness at the time, and without memory of it afterwards.' What is here said of delusions in that extreme form in which they assume unmistakeably the character of madness applies equally, as an explanation, to those less obtrusive, though far more frequent, forms in which they produce semi-insanity, monomania, melancholy, or partial and temporary deception. In all these cases, the

[20] Maudsley, *The Physiology and Pathology of Mind*, ch. iv. p. 101.

sensations are really experienced; the error consists in referring the cause of the sensations to external objects rather than to the morbid condition or action of the brain itself. The testimony of others, or the inherent improbability of the things perceived, ought to be regarded, though they seldom are, as sufficient proof that the evidence of the senses is given under abnormal and untrustworthy conditions.

To the head of Mal-observation, or the substitution of gratuitous inference for accurate observation, may be referred the fallacy which may, perhaps, best be designated as *Exaggerated Comparison.* By the side of anything very large, we are apt, being prepossessed by the idea of largeness, to suppose that a small object is smaller than it really is, and, on the other hand, by the side of anything very small, being prepossessed by the idea of smallness, that a large object is larger than it really is. Similarly, of things bright or dark, of periods and distances long or short, of actions good or evil, of evidence probable or improbable, and the like. We are all familiar with the 'only half a mile off,' when we are approaching a town, which we shall probably find to be at least double or treble the distance named. The countryman, of whom we enquire, knows that, in comparison with long distances, the distance is short, and then, unconsciously exaggerating the shortness of the distance, proceeds to name some definite distance which, in his mind, is typical of shortness or which, he thinks, will sufficiently reassure the weary traveller. Instead of recurring to his own actual experience of the time he takes to walk it, he draws a rapid inference from the fact that the distance is comparatively short to the particular distance which he names. In the same way, men are apt to underrate the probability of an argument as compared with certainty or a very high degree of probability, or, on the other hand, to overrate its probability, as compared with very faint indications of evidence. The concentration of our attention on one term of the comparison perverts our judgment with reference to the other term.

The description here given of the errors originating in Non-observation or Mal-observation includes, as will already have been perceived, the errors incident to artificial as well as to natural observation, that is, to experiment as well as to observation proper.

III. The errors incidental to the other operations preliminary to induction, namely, classification, nomenclature, terminology, and hypothesis, will be sufficiently apparent on a perusal of the sections appropriated to the discussion of those processes. In the steps intermediate between the observation of individual facts and the inductive inference itself, it is in the employment of artificial instead of natural classifications, and in the neglect of the rules designed to guard against the formation of illegitimate hypotheses, that the danger of error mainly lies.

B. The fallacies incidental to the performance of the inductive process itself may be called Fallacies of Generalisation: An error of this class is committed whenever, in arguments grounded on experience, we overrate the value of the evidence before us; that is, whenever we accept an imperfect induction as a perfect one, or whenever, in an induction confessedly imperfect, we underestimate the amount of imperfection.

Of the imperfect inductions, the argument from analogy is little likely to be mistaken for a perfect induction. The strength of the analogy is often grossly exaggerated, and an argument which possesses little or no probability is often adduced as affording highly probable evidence; but, as this kind of argument is very seldom [21] treated as affording absolute certainty, the discussion of false analogies may be reserved till I have completed the treatment of those errors which consist in regarding imperfect as perfect inductions.

Excluding analogy, there are, as we have seen, two forms of imperfect induction, that which employs the incomplete Inductio per Enumerationem Simplicem and that which consists in an imperfect fulfilment of the conditions of the inductive· methods. An argument of either of these classes may be, and frequently is, mistaken for a perfect induction. I shall first

---

[21] The geological example on p. 236 may perhaps be an instance of an analogical argument thus regarded. Many writers have certainly treated the inference as if its certainty admitted of no doubt.

notice the case in which scientific induction is simulated by the incomplete Inductio per Enumerationem Simplicem [22].

IV. When men first begin to argue from their experience of the past to their expectation of the future, or from the observation of what immediately surrounds them to the properties of distant objects, they seem naturally to fall into this unscientific and unreflective mode of reasoning. They have constantly seen two phenomena in conjunction, and, consequently, they cannot imagine them to be dissevered, or they have never seen two phenomena in conjunction, and, consequently, they cannot imagine them to be associated. The difficulties experienced by children in accommodating their conceptions to the wider experiences of men; the tendency of the uninstructed, and frequently even of the instructed, to invest with the peculiar circumstances of their own time or country the men of a former generation or of another land; the prejudices entertained against those of another creed, or party, or nationality, as if moral excellence were never dissociated from particular opinions or a particular lineage, — are all evidences of the limited character of our first efforts at generalisation. It is long before men learn to discriminate between the

[22] The student who has read the first and fourth Chapters hardly needs to be reminded that there are cases, however, in which the method of Inductio per Enumerationem Simplicem may, or even must, be employed. The fallacies here treated are due to the unnecessary or injudicious employment of the method.

material and immaterial circumstances attendant on
any given phenòmenon, to perceive the irrelevancy of
the immaterial circumstances, and to recognise the
necessity of insisting on a repetition of all the material
circumstances before they anticipate a similar effect.
But not only is the Inductio per Enumerationem Simplicem the mode of generalisation natural to immature
and uninstructed minds; it is the method which, till the
time of Bacon [23], or at least till the era of those great
discoveries which shortly preceded the time of Bacon,
was almost universal. Aristotle, it is true, usually [24] (for

[23] Bacon seems to be never weary of condemning this unscientific
procedure. Thus, in addition to the aphorism already quoted (p.
125), we have, amongst others, the following emphatic passages:
'Axiomata quæ in usu sunt ex tenui et manipulari experientia, et
paucis particularibus, quæ ut plurimum occurrunt, fluxere; et sunt
fere ad mensuram eorum facta et extensa: ut nil mirum sit, si ad
nova particularia non ducant. Quod si forte instantia aliqua, non
prius animadversa aut cognita, se offerat, axioma distinctione aliqua
frivola salvatur, ubi emendari ipsum verius foret.'—*Nov. Org.* Lib. I.
Aph. xxv. 'At philosophiæ genus *empiricum* placita magis deformia
et monstrosa educit, quam *sophisticum* aut rationale genus; quia non
in luce notionum vulgarium (quæ licet tenuis sit et superficialis,
tamen est quodammodo universalis, et ad multa pertinens) sed in
paucorum experimentorum angustiis et obscuritate fundatum est. . . .
Sed tamen circa hujusmodi philosophias cautio nullo modo præter-
mittenda erat; quia mente jam prævidemus et auguramur, si quando
homines, nostris monitis excitati, ad experientiam se serio contulerint
(valere jussis doctrinis sophisticis), tum demum, propter præmaturam
et præproperam intellectus festinationem et saltum sive volatum ad
generalia et rerum principia, fore ut magnum ab hujusmodi philosophiis periculum immineat; cui malo etiam nunc obviam ire
debemus.'—Aph. lxiv.

[24] For exceptions, see *An. Post.* I. 31, p. 88, a. 4–5 ; Top. VIII. 8,

he is not consistent on this point) requires that an induction should be based on an examination of all the instances; but this requirement being in the vast majority of cases (even if we suppose Aristotle to be speaking of species rather than individuals) impossible of fulfilment, he was obliged, whenever he had recourse to experience, to content himself with an inspection of those cases which were nearest at hand. Thus, in the very passage [25] in which he emphatically asserts that the minor premiss of the inductive syllogism (for he represents induction under the syllogistic form) should include all the instances, he argues that all animals which are deficient in bile are long-lived, because he finds this to be the case with the man, the horse, and the mule. Aristotle's works, and especially those on Natural History, abound in rash generalisations of this kind. 'It is a fact,' says Mr. Lewes [26], 'that normally in turtles, and exceptionally in elephants, horses, and oxen, there is an ossification of the septum of the heart. Aristotle saw or heard of one of these "bones" in the hearts of a horse and an ox, and forthwith generalised the observation

p. 160, b. 3. For further information on Aristotle's theory of Induction, I must refer the student to the beginning of § 13, and, for a more detailed account of the causes of his failure in his physical researches than can well be given here, to § 11 of the Introduction to my edition of Bacon's *Novum Organum*.

[25] *Analytica Priora*, ii. 25.

[26] Lewes' *Aristotle*, ch. xvi. § 399. On the other hand, the student, who is interested in the history of science, will do well to read, in arrest of judgment on Aristotle, Dr. William Ogle's Introduction to his translation of the *De Partibus Animalium*.

thus : "The heart is destitute of bones except in horses and in a species of ox; these, however, in consequence of their size, have something bony as a *support*, just as we find throughout the whole body[27]." His Spanish follower Funes Y Mendoça improves on this statement by saying that the bone acts like a stick to support the weight of the heart, which is very great.'

There is another passage in which Aristotle tells us that the cranium of a dog consists of a single bone[28]. 'It is probable,' says the author of the review previously quoted[29], 'that Aristotle had got hold of the cranium of an old individual in which the sutures had become obliterated.'

The employment of the Inductio per Enumerationem Simplicem prevailed so universally from the time of Aristotle to the rise of modern science that it seems unnecessary to multiply instances of it during that period. But it may be instructive to illustrate from the history of more recent times the peculiar facility with which some even of the greatest discoverers have lapsed into this erroneous form of reasoning.

'Bichat,' says Mr. Lewes[30], 'tried to establish a generalisation which has been much admired, namely, that all the

---

[27] *De Partibus Animalium*, Bk. III. ch. iv.

[28] τὰ μὲν γὰρ ἔχει μονόστεον τὸ κρανίον, ὥσπερ ὁ κύων, τὰ δὲ συγκείμενον, ὥσπερ ἄνθρωπος.—*Historia Animalium*, Bk. III. ch. vii.

[29] *Quarterly Review*, No. 233, Art. ii. The mention of the human skull, which had no sutures, is evidently borrowed from Herodotus, IX. 83.

[30] Lewes' *Aristotle*, ch. xvi. § 399 d.

organs of Animal life are double and symmetrical, while all the organs of Vegetal life are single and asymmetrical. Unhappily the facts do not fit.   In the commencement almost *every* organ is double and symmetrical ; and only in the later stages of development do the differences appear. Even in the matured organism we find many striking exceptions to Bichat's generalisation.   Thus the parotid, sublingual, and mammary glands, the lungs, the kidneys, ovaries, and testes, are all vegetal organs, and all generally double.   And if the heart and uterus are classed as single organs, then must the brain and spinal cord be classed thus. While in birds the liver is double and symmetrical.'

'It is in a great degree true,' we are informed by Dr. Paris[31], ' that the sensible qualities of plants, such as *colour, taste,* and *smell,* have an intimate relation to their properties, and may often lead by analogy to an indication of their powers ; we have an example of this in the dark and gloomy aspect of the *Luridæ,* which is indicative of their narcotic and very dangerous qualities, as *Datura, Hyoscyamus, Atropa,* and *Nicotiana. Colour* is certainly in many cases a test of activity; the deepest of coloured flowers, the *Digitalis,* for example, are the most active, and when the leaves of powerful plants lose their green hue, we may conclude that a corresponding deterioration has taken place with respect to their virtues.: but Linnæus ascribed too much importance to such an indication, and his aphorisms are unsupported by facts ; for instance, he says, " Color pallidus *insipidum,* viridis *crudum,* luteus *amarum,* ruber *acidum,* albus *dulce,* niger *ingratum,* indicat." '

The early history of Geology presents, in the controversy which was long carried on between the Neptunians and Vulcanians, a remarkable instance of the

[31] *Pharmacologia,* ninth ed. pp. 110, 111.

errors arising from a partial induction, as well as of the tenacity with which men will cling to views to which they have once committed themselves. The Neptunians, the student need hardly be told, referred all geological phenomena to the influence of water, while the Vulcanians greatly exaggerated the action of heat in the past history of the globe, and multiplied to an excess the number of formations to be ascribed to an igneous origin. Of the Neptunians, the great Saxon geologist Werner was the chief.

'Werner,' says Sir Charles Lyell[32], 'had not travelled to distant countries; he had merely explored a small portion of Germany, and conceived, and persuaded others to believe, that the whole surface of our planet, and all the mountain chains in the world, were made after the model of his own province. It became a ruling object of ambition in the minds of his pupils to confirm the generalisations of their great master, and to discover in the most distant parts of the globe his "universal formations," which he supposed had been each in succession simultaneously precipitated over the whole earth from a common menstruum or "chaotic fluid." It now appears that the Saxon professor had misinterpreted many of the most important appearances even in the immediate neighbourhood of Freyberg. Thus, for example, within a day's journey of his school, the porphyry, called by him primitive, has been found not only to send forth veins or dikes through strata of the coal formation, but to overlie them in mass.'

'In regard to basalt and other igneous rocks, Werner's theory was original, but it was also extremely erroneous.

[32] Lyell's *Principles of Geology*, ninth ed. Bk. I. ch. iv.

The basalts of Saxony and Hesse, to which his observations were chiefly confined, consisted of tabular masses capping the hills, and not connected with the levels of existing valleys, like many in Auvergne and the Vivarais. These basalts, and all other rocks of the same family in other countries, were, according to him, chemical precipitates from water. He denied that they were the products of submarine volcanoes; and even taught that, in the primeval ages of the world, there were no volcanoes.'

After describing the complete demolition of this theory by some of Werner's contemporaries, Sir Charles Lyell adds :—

'Notwithstanding this mass of evidence, the scholars of Werner were prepared to support his opinions to their utmost extent; maintaining, in the fulness of their faith, that even obsidian was an aqueous precipitate. As they were blinded by their veneration for the great teacher, they were impatient of opposition, and soon imbibed the spirit of a faction; and their opponents, the Vulcanists, were not long in becoming contaminated with the same intemperate zeal. Ridicule and irony were weapons more frequently employed than argument by the rival sects, till at last the controversy was carried on with a degree of bitterness almost unprecedented in questions of physical science. Desmarest alone, who had long before provided ample materials for refuting such a theory, kept aloof from the strife; and, whenever a zealous Neptunist wished to draw the old man into an argument, he was satisfied with replying "Go and see."'

In the Science of Probability, there is an interesting example of the unreflecting application of the Inductio per Enumerationem Simplicem. Averages of a sufficiently trustworthy character can often be struck as to the frequency of such events as the number of deaths, the number of suicides, the number of lost letters which occur in a year. But the least reflexion ought to show that the accuracy of these calculations depends on the assumption that the causes in operation, so far as they affect these events, will continue to be much the same as at present. This, however, is a consideration which is frequently lost sight of, and thus averages, which may be perfectly true within certain limits and on certain hypotheses, are extended, as if they were true universally and unconditionally. Mr. Venn, in his work on the *Logic of Chance*[33], has drawn especial attention to this source of error. The following passage selected from that work will, perhaps, afford a sufficient illustration of the point in question :—

'Let us take, for example, the average duration of life. This, provided our data are sufficiently extensive, is known to be tolerably regular and uniform. But a very little consideration will show that there may be a superior as well as an inferior limit to the extent within which this uniformity can be observed. At the present time the average duration of life in England may be, say thirty; but a century ago it was decidedly less; several centuries ago it was very much less; whilst, if we possessed statistics referring to our early British ancestors, we should probably find that there has been since that time a still more marked improvement. What may be the future tendency no man can say for certain. It may be, and we hope will be the case, that, owing to sanitary

---

[33] Venn's *Logic of Chance*, chap. i.

and other improvements, the duration of life will go on increasing steadily ; it is quite conceivable that it should do so without limit. On the other hand, this duration might gradually tend towards some fixed length. Or, again, it is perfectly possible that future generations might prefer a short and a merry life, and therefore reduce their average. All that I am concerned to indicate is, that this uniformity (as we have hitherto called it) has varied, and, under the influence of future eddies in opinion and practice, may vary still ; and this to any extent, and with any degree of irregularity. To borrow a term from Astronomy, we find our uniformity subject to what might be called an irregular *secular* variation.

'The above is a fair typical instance. If we had taken a less simple feature than the length of life, or one less closely connected with what may be called the great permanent uniformities of nature, we should have found the peculiarity under notice exhibited in a far more striking degree. The deaths from small-pox, for example, or the instances of duelling or accusations of witchcraft, if examined during a few successive years, would have shown a very tolerable degree of uniformity. But this uniformity has risen probably from zero ; after various and very great fluctuations seems tending towards zero again ; and may, for anything we know, undergo still greater fluctuations in future. Now these examples I consider to be only extreme ones, and not such very extreme ones, of what is the almost universal rule in nature. I shall endeavour to shew that even the few apparent exceptions, such as the proportions between male and female births, &c., may not be, and probably in reality are not, exceptions. A type that is persistent and invariable is scarcely to be found in nature[34].'

In these and similar cases, the fallacy arises from

[34] Venn's *Logic of Chance*, ch. i. sect. 10, 11.

supposing that mere frequency of occurrence affords a sufficient guide to inference, without reflecting that the events depend on causes, and that, if the causes vary, the character of the events must vary with them.

Sometimes, frequency of occurrence, instead of furnishing an argument for the recurrence of an event, ought, if we duly reflect on the natural action of causes, actually to furnish an argument against it. Thus, a miner, instead of trusting to his rope, because it has served him so often, ought actually to distrust it, because it has been strained so much; a prodigal, who has frequently succeeded in borrowing from his friends, ought to begin to suspect that their patience may be exhausted; a timid man, who has on one or two occasions aroused his neighbours by a false alarm, instead of arguing from experience that they will come to his rescue again, ought rather to expect that, warned by the past, they will remain comfortably in their beds. It cannot be too often repeated, that we ought never to depend on frequency of occurrence, wherever it is possible to have recourse to facts of causation.

It is remarked by Mr. Mill that the Method of Simple Enumeration, though almost banished from the physical sciences, is still the common and received method of induction in whatever relates to man and society. The reason of this remark is to be sought in the extraordinary difficulty of subjecting this class of speculations to the more scientific methods. Moral and social phenomena are so complex that it is often next to impossible to discover by elimination the true connexion between any two events or sets of facts. Take, for instance, such questions as the influence of any particular form of government upon the welfare of the people among whom it is established, the effects of religion, or of any particular form of religion, upon morals, the social and political conditions most favourable to the development of art or literature or science or commerce. Here, if it be required to discover the cause of a given effect, our materials are a set of consequents constantly varying in their character and intensity, and a set of antecedents, often very numerous, any one of which may have an appreciable influence in the production of the effect in question; and it is obvious that to detect the precise degree in which the effect is due to any one of these antecedents, even supposing the task to be possible, will require the utmost skill, patience, and dispassionateness in the selection and comparison of instances. Nor, if it be required to discover the effect of a given cause, will the task be much simplified; for, though it may be possible to fix the precise time at which a new cause— say a new form of religion, a new form of government, or a new commercial tariff—was introduced, yet, before it can be argued that any novel event which may appear to have resulted from it, is really due to it, as an effect to a cause, the enquirer is bound to satisfy himself (1) that the introduction of the new cause was not accompanied by other causes which may have wholly or partially produced the supposed effect, (2) that the new cause and the supposed effect are not joint effects of some common cause which he may have overlooked. It is the extreme difficulty of bringing this class of questions within the requirements of scientific induction, that has led, on the one hand, to the employment of the loose Method of Inductio per Enumerationem Simplicem, or of a mere appeal to unsifted experience, and on the other to the disbelief in the possibility of arriving at any satisfactory conclusions upon them. At the same time, there can be little doubt that moral and social enquiries are beginning to emerge from the chaotic state of confusion in which they have hitherto been sunk, and that what are now dignified with the titles of the moral and political sciences, however imperfect they may be, are beginning to be something more than mere collections of random guesses, or conclusions drawn from the first undisciplined impressions of the teaching of experience.

To the class of fallacies originating in the employment of the incomplete Inductio per Enumerationem Simplicem may perhaps be referred the illegitimate use of the Argument from Authority. The opinions or predictions of a certain man or of a certain class of men upon some particular question or questions have been subsequently found to be verified by the issue of events or an examination of the facts. From this circumstance it is sufficiently rash to infer, without further warrant, that the correspondence between these predictions or opinions and the subsequent events or ascertained facts is the result of knowledge, and not of what we call

accident; but, not content even with this inference, men are apt to draw the far more unwarrantable one that this person or class of persons is to be accepted as an authority on all matters, or at least on all matters of the same or of an analogous kind. It is on this principle that a savage, or even an uneducated man in a civilised community, will trust implicitly any person for whom he has conceived a general respect. In nine cases out of ten he probably acts more wisely in trusting to such a person than in trusting to himself. But the same habit of mind, which is a virtue among uneducated men and in primitive states of society, becomes one of the most serious obstacles to progress and knowledge when men, either individually or collectively, have at-tained that stage at which they are able to enquire for themselves. We have to learn not only that men are to be trusted exclusively within the limits of their own experience, in their own profession or pursuit, but that even within those limits their authority is apt to become tyrannical and irrational unless it is constantly con-fronted with facts and subjected to the criticism of others.

But an undiscriminating submission to the authority of contemporaries, of which I have hitherto exclusively spoken, has been but a slight source of error when com-pared with undiscriminating submission to the authority of past generations [35]. The latter involves a kind of com-

[35] Of this tendency we have many 'glaring instances,' as Bacon would call them. The error has been, so to say, canonised in the

pound fallacy. The authority of an Aristotle or a Galen has come, by the process already described, to be received without question and without limit by his own or by the succeeding generation; and then, by the constant repetition of a similar process, it is received from that generation by the leading minds of the next, from them by their contemporaries, and so on, respect for tradition being blended with respect for a great name, and both these resting for their support on the deference paid to established authority. Many of the propositions accepted without the slightest hesitation by previous generations on this kind of authority now appear to us patently absurd, nor is it without effort that we can realise the universality of their former reception [36]. Instances of such propositions have already

proverb 'Mallem cum Platone errare.' There is a characteristic anecdote of Scheiner, who contests with Galileo the honour of having been the first to observe the spots in the sun. 'Scheiner was a monk; and, on communicating to the superior of his order the account of the spots, he received in reply from that learned father a solemn admonition against such heretical notions :—" I have searched through Aristotle," he said, " and can find nothing of the kind mentioned: be assured, therefore, that it is a deception of your senses, or of your glasses." '—Baden Powell's *History of Natural Philosophy*, p. 171.

[36] The increasing unwillingness of men to accept a proposition on mere authority is thus forcibly put by Bentham, *Book of Fallacies*, Part I. ch. i., first published in French by M. Dumont, in 1815, and in English by ' A Friend,' in 1824.

' As the world grows older, if at the same time it grows wiser (which it will do, unless the period shall have arrived at which experience, the mother of wisdom, shall have become barren), the been given under the head of the Fallacies of Non-Observation, to the production of which class of fallacies the undue devotion to authority has, perhaps, contributed more than any other cause[37]. But, in subjects lying remote from ordinary observation, propositions almost equally absurd have held their ground till quite recently; some continue still to maintain themselves, and others no doubt will be propounded, from time to time, to take advantage of the credulity of mankind.

'To give a general currency,' says Dr. Paris[38], 'to a hypo-

influence of authority will in each situation, and particularly in parliament, become less and less.'

'Take any part of the field of moral science, private morality, constitutional law, private law; go back a few centuries, and you will find argument consisting of reference to authority, not exclusively, but in as large a proportion as possible. As experience has increased, authority has been gradually set aside, and reasoning, drawn from facts, and guided by reference to the end in view, true or false, has taken its place.

*   *   *   *   *   *

'In mechanics, in astronomy, in mathematics, in the new-born science of chemistry—no one has at this time of day either effrontery or folly enough to avow, or so much as to insinuate, that the most desirable state of these branches of useful knowledge, the most rational and eligible course, is to substitute decision on the ground of authority to decision on the ground of direct and specific evidence.'.

[37] It might appear that the illegitimate use of the Argument from Authority should be classed amongst the Fallacies of Non-Observation; but, though a blind devotion to authority is one of the most powerful influences in leading men to neglect observation and experiment, the disposition to bow thus unduly to it is itself a fact which requires explanation, and one which it is here attempted to explain.

[38] Dr. Paris' *Pharmacologia*, Introduction, p. 76, &c.

thetical opinion, or medicinal reputation to an inert substance, nothing more is required than the talismanic aid of a
few great names; when once established upon such a basis,
ingenuity, argument, and even experiment, may open their
ineffectual batteries; the laconic sentiment of the Roman
satirist is ever opposed to remonstrance :—"*Marcus dixit?
ita est.*" A physician cannot err in the opinion of the
public, if he implicitly obeys the dogmas of authority. In
the most barbarous ages of ancient Egypt, he was pun-
ished or rewarded according to the extent of his success;
but to escape the former it was only necessary to show that
an orthodox plan of cure had been followed, such as was
prescribed in the acknowledged writings of Hermes. It is
an instinct in our nature to follow the track pointed out by
a few leaders; we are gregarious animals, in a moral as well
as a physical sense, and we are addicted to routine because
it is always easier to follow the opinions of others than to
reason and judge for ourselves; and thus do one half of the
world live as alms-folk on the opinions of the other half.
What but such a temper could have upheld the preposterous
system of Galen for more than thirteen centuries, and have
enabled it to give universal laws in medicine to Europe,
Africa, and part of Asia? What but the spell of authority
could have inspired a general belief that the sooty washings
of resin could act as a universal remedy? What but a blind
devotion to authority, or an insuperable attachment to established custom and routine, could have so long preserved from
oblivion the absurd medicines which abound in our earlier
dispensatories? for example, the "*Decoctum ad Ictericos*" of
the Edinburgh College, which never had any foundation but
that of the doctrine of signatures in favour of the *Curcuma*
and *Chelidonium majus;* and it is only within a few years
that the *Theriaca Andromachi*, in its ancient form, has been
dismissed from our Pharmacopœia. The CODEX MEDICA-
MENTARIUS of Paris still cherishes the many-headed monster

of pharmacy, under the appropriate title of " *Electuarium Opiatum Polypharmacum.*"'

'The same devotion to authority which induces us to retain an accustomed remedy with pertinacity, will frequently oppose the introduction of a novel practice with asperity, unless indeed it be supported by authority of still greater weight and consideration. The history of various articles of diet and medicine will prove in a striking manner how greatly their reputation and fate have depended upon authority. It was not until many years after *Ipecacuan* had been imported into Europe, that Helvetius, under the patronage of Louis XIV, succeeded in introducing it into practice : and to the eulogy of Katharine, queen of Charles II, we are indebted for the general introduction of tea into England.'

'The history of the warm bath presents us with another curious instance of the vicissitudes to which the reputation of our valuable resources is so universally exposed ; that which for so many ages was esteemed the greatest luxury in health, and the most efficacious remedy in disease, fell into total disrepute in the reign of Augustus, for no other reason than because Antonius Musa had cured the emperor of a dangerous malady by the use of the *cold* bath. The most frigid water that could be procured was, in consequence, recommended on every occasion : thus Horace, in his epistle to Vala, exclaims—

"Caput ac stomachum supponere fontibus audent
Clusinis, Gabiosque petunt, et frigida rura."—*Epist.* xv. lib. i.

'This practice, however, was doomed but to an ephemeral popularity, for, although it had restored the emperor to health, it shortly afterwards killed his nephew and son-in-law, Marcellus ; an event which at once deprived the remedy of its credit and the physician of his popularity.

'The history of the Peruvian bark would furnish a very curious illustration of the overbearing influence of authority in giving celebrity to a medicine, or in depriving it of that reputation to which its virtues entitle it. This heroic remedy was first brought to Spain in the year 1632, and we learn from Villerobel that it remained for seven years in that country before any trial was made of its powers, a certain ecclesiastic of Alcala being the first person in Spain to whom it was administered in the year 1639; but even at this period its use was limited, and it would have sunk into oblivion but for the supreme power of the Roman church, by whose auspices it was enabled to gain a temporary triumph over the passions and prejudices which opposed its introduction. Innocent the Tenth, at the intercession of Cardinal de Lugo, who was formerly a Spanish Jesuit, ordered that the nature and effects of it should be duly examined, and, upon being reported as both innocent and salutary, it immediately rose into public notice; its career, however, was suddenly stopped by its having unfortunately failed, in the autumn of 1652, to cure Leopold, Archduke of Austria, of a quartan intermittent; this disappointment kindled the resentment of the prince's principal physician, Chifletius, who published a violent philippic against the virtues of Peruvian bark, which so fomented the prejudices against its use, that it had nearly fallen into total neglect and disrepute.'

In discussing the *Argument from Authority*, I have already touched on the *Argument from Universal Consent.* 'This is a proposition to which we cannot refuse our assent, for it is accepted by all mankind.' In dealing with this argument, we must always ask, first of all, whether the proposition assented to expresses an immediate perception or an inference. If it expresses the former, we cannot call it in question, for the immediate perceptions of men are ultimate facts, true, at all events, to us, and admitting of no further test. But if the proposition expresses an inference, as, for instance, in the case of the belief in the motion of the sun round the earth, or the non-existence of antipodes, we must proceed to ask further what are the grounds of the inference, and, unless the grounds of the inference approve themselves to us, we are at liberty to doubt or reject it. At the same time, this argument, even though the proposition only express an inference, may possess considerable, if not overwhelming, force, provided that the conclusion has been arrived at by a number of competent persons after due examination, and as a result of independent investigation. Even here, however, the true authority is that of the competent investigators, not that of their credulous or incompetent followers [39]. The latter, as was once said by the late Bishop Thirlwall, may be regarded as the ciphers after a decimal point [40].

V. The errors incident to the employment of the various Inductive Methods have already been pointed out

[39] ' Verus enim consensus is est, qui ex libertate judicii (re prius explorata) in idem conveniente consistit. At numerus longe maximus eorum, qui in Aristotelis philosophiam consenserunt, ex præjudicio et auctoritate aliorum se illi mancipavit; ut sequacitas sit potius et coitio, quam consensus.'—Bacon, *Nov. Org.*, Lib. I. Aph. lxxvii.

[40] Cp. Glanvill's *Scepsis Scientifica*, ch. xvii.: ' Authorities alone with me make no number, unless Evidence of Reason stand before them: for all the Cyphers of Arithmetic are no better than a single nothing.'

in my detailed description of each of these Methods, but it may be useful in this place to take note of certain forms of fallacy which appear to be common to them all.

The Inductive Methods may all be regarded as devices for the elimination of extraneous circumstances and for the establishment of a causal connexion between some two phenomena, a and b, the connexion which it is sought to establish being generally that of cause and effect. Now, in our investigation, we may either have mistaken the precise relation between a and b, or we may have overlooked some other material circumstance or group of circumstances, c. In the former case, the most common sources of error are either the inversion of cause and effect or the neglect of their reciprocal action, the 'mutuality of cause and effect,' as it is called by Sir G. C. Lewis. In the latter case (supposing a to be the presumed cause, and b the presumed effect), it seems open to us to have committed any of the following errors: (1) to have mistaken a for the cause, when the real cause is c; (2) to have mistaken a for the sole cause, when a and c are the joint causes, either (a) as both contributing to the *total* effect, or (β) as being both essential to the production of any effect whatever[41]; (3) to have mistaken a for the cause of b, when they are really

[41] The distinction may be illustrated by a familiar example. If a cistern is filled by two pipes, the water passing through each *contributes* to the *total* amount of water in the cistern. But, if the cistern is filled by one pipe having two taps, one above the other, both taps must be turned in order that the cistern may receive any water whatever.

both of them effects of c; (4) to have mistaken a for the proximate cause of b, when it is really only the remote cause, c, which has escaped our attention, being the proximate cause.

To begin with the latter class of errors.

(1) The following extract from Mr. Lewes' *Physiology of Common Life*[42] may serve as an illustration of the first subdivision :—

'One very general, indeed almost universal, misconception on this subject (asphyxia or suffocation) is that carbonic acid is poisonous in the blood ; but the truth seems to be that the carbonic acid is noxious only when it prevents the access of oxygen. There is always carbonic acid in the blood, both venous and arterial. Its accumulation in the blood is only fatal when there is such an accumulation *in the atmosphere* as will prevent its exhalation ; its mere presence in the blood seems to be quite harmless, even in large quantities, provided always that it be not retained there to the exclusion of oxygen. Carbonic acid, when absorbed into the blood, which is alkaline, cannot there exert its irritant action as an acid, because it will either be transformed into a carbonate or be dissolved. Bernard has injected large quantities into the veins and arteries, and under the skin, of rabbits, and found no noxious effect ensue. The more carbonic acid there is in the blood, the more will be exhaled, provided always that the *air* be not already so charged with it as to prevent this exhalation.'

Here there are really two antecedents, the presence of carbonic acid and the exclusion of oxygen, and the noxious effects, which are erroneously ascribed to the

[42] Vol. i. p. 383.

former cause, ought properly to be referred to the latter.

The above extract exemplifies this error as vitiating an application of the Method of Agreement. In the following extracts from Dr. Paris' *Pharmacologia*, it will be seen also to vitiate applications of the Method of Difference :—

'Soranus, who was contemporary with Galen, and wrote the life of Hippocrates, tells us that honey proved an easy remedy for the aphthæ of children ; but, instead of at once referring the fact to the medical qualities of the honey, he very gravely explains it, from its having been taken from bees that hived near the tomb of Hippocrates[43] ! '

'In my life of Sir Humphry Davy, I have published an anecdote which was communicated to me by the late Mr, Coleridge, and which bears so strikingly upon the present subject that I must be excused for repeating it. As soon as the powers of *nitrous oxide* were discovered, Dr. Beddoes at once concluded that it must necessarily be a specific for paralysis: a patient was selected for the trial, and the management of it was entrusted to Davy. Previous to the administration of the gas, he inserted a small pocket thermometer under the tongue of the patient, as he was accustomed to do upon such occasions, to ascertain the degree of animal temperature, with a view to future comparison. The paralytic man, wholly ignorant of the nature of the process to which he was to submit, but deeply impressed, from the representations of Dr. Beddoes, with the certainty of its success, no sooner felt the thermometer under his tongue, than he concluded the *talisman* was in full operation, and in a burst of

[43] *Pharmacologia,* p. 20.

enthusiasm declared that he already experienced the effect of its benign influence throughout his whole body : the opportunity was too tempting to be lost ; Davy cast an intelligent glance at Mr. Coleridge, and desired his patient to renew his visit on the following day, when the same ceremony was performed, and repeated every succeeding day for a fortnight, the patient gradually improving during that period, when he was dismissed as cured, no other application having been used[44]'

'Amongst the numerous instances which have been cited to show the power of faith over disease, or of the mind over the bodily organs, *the cures performed by royal touch* have been considered the most extraordinary : but it would appear, upon the authority of Wiseman, that the cures which were thus effected were in reality produced by a very different cause ; for he states that part of the duty of the royal physicians and serjeant surgeons was to select such patients afflicted with scrofula as evinced a tendency towards recovery, and that they took especial care to choose those who approached the age of puberty. In short, those only were produced whom Nature had shown a disposition to cure ; and as the touch of the king, like the sympathetic powder of Digby, secured the patient from the mischievous importunities of art, so were the efforts of Nature left free and uncontrolled, and the cure of the disease was not retarded or opposed by the administration of adverse remedies. The wonderful cures of Valentine Greatricks, performed in 1666, which were witnessed by contemporary prelates, members of parliament, and fellows of the Royal Society, amongst whom was the celebrated Mr. Boyle, would probably, upon investigation, admit of a similar explanation. It deserves, however, to be noticed that, in all records of extraordinary cures performed by mysterious agents, there has

[44] *Pharmacologia,* p. 28.

always been a desire to conceal the remedies and other curative means which might have been simultaneously administered. Thus Oribasius commends, in high terms, a necklace of *peony-root* for the cure of epilepsy ; but we learn that he always took care to accompany its use with copious evacuations, although he assigns to them not the least share of credit in the cure. In later times, we have an excellent specimen of this species of deception, presented to us in a work on scrofula by Mr. Morley, written, as we were informed, for the sole purpose of restoring the much-injured character and use of the *vervain ;* in which the author directs the root of that plant to be tied with *a yard of white satin riband* around the neck ;—but mark—during the period of its application, he calls to his aid the most active medicines in the materia medica. " It is unquestionable," says Voltaire, speaking of sorceries, "that certain words and ceremonies will effectually destroy a flock of sheep, if administered with a sufficient portion of arsenic[45]." '

' Our inability upon all occasions to appreciate the efforts of nature, in the cure of disease, must necessarily render our notions, with respect to the powers of art, liable to numerous errors and deceptions. Hence protracted or *wire-drawn* cures ought to be very cautiously received as evidences of the success of medical treatment. Many diseases require only time to enable nature to remove them. All the long train connected with hysteria are *cured by time ;* the solution of which, as Mr. Travers has observed, is to be found in the fact that the hysteric period wanes, and the restlessness of the temperament undergoes a slow but salutary change. Nothing, certainly, is more natural, although it may be very erroneous, than to attribute the cure of a disease to the last medicine that had been administered ; the advocates even of amulets and charms have been thus enabled to appeal to

[45] *Pharmacologia,* p. 30.

the testimony of what they call experience, in justification of their superstition[46].'

Of a similar character was the old superstition, noticed by Sir Thomas Browne[47] and many other authors, that the hardest stone could be broken by goat's blood :—

'And, first, we hear it in every mouth, and in many good authors read it, that a diamond, which is the hardest of stones, not yielding unto steel, emery, or any thing but its own powder, is yet made soft, or broke by the blood of a goat. . . . . But this, I perceive, is easier affirmed than proved. For lapidaries, and such as profess the art of cutting this stone, do generally deny it; and they that seem to countenance it have in their deliveries so qualified it, that little from thence of moment can be inferred for it. For first, the holy fathers, without a further enquiry, did take it for granted, and rested upon the authority of the first deliverers. . . . But the words of Pliny, from whom most likely the rest at first derived it, if strictly considered, do rather overthrow, than any way advantage this effect. His words are these : *Hircino rumpitur sanguine, nec aliter quàm recenti calidoque macerata, et sic quoque multis ictibus, tunc etiam præterquam eximias incudes malleosque ferreos frangens.* That is, it is broken with goat's blood, but not except it be fresh and warm, and that not without many blows, and then also it will break the best anvils and hammers of iron.'

The example of Sir Kenelm Digby's sympathetic powder (already quoted pp. 270–1) also illustrates this class of fallacies[48].

[46] *Pharmacologia*, p. 88.

[47] *Enquiry into Vulgar and Common Errors*, Bk. II. ch. v. Collected Works, vol. ii. pp. 334, 335.

[48] These instances, together with many others in this chapter,

It should be noticed that, when we attribute a phenomenon to a wrong cause, it does not always follow that
this cause, had it been in action, might not have produced the event. Thus, we may wrongly attribute death
in some given case to poison, or infection to actual
contact with a diseased person, or ignition to friction,
because these causes were not then and there in action,
though, had they been actually operating, they would
have been perfectly competent to produce the effect.
When we make a mistake of this kind, it frequently arises
from our concentrating our attention exclusively on some
one or a few of the possible causes which may produce
a given effect, thus neglecting the consideration of the
Plurality of Causes, to which attention has repeatedly
been drawn in the previous pages [49].

(2) When an effect is the joint result of two or more

illustrate the ancient fallacies ' Non causa pro causa,' and ' Post hoc,
ergo propter hoc.' It will probably have already occurred to the
student that some of the examples just cited might have been equally
well adduced as examples of the fallacy of non-observation. It, in
fact, frequently happens that the same error may be assigned indifferently to two or more sources of deception. 'From the elliptical
form,' says Archbishop Whately (*Elements of Logic,* Bk. iii. § 1),
' in which all reasoning is usually expressed, and the peculiarly
involved and oblique form in which fallacy is for the most part
conveyed, it must of course be often a matter of doubt, or rather of
arbitrary choice, not only to which genus each *kind* of fallacy should
be referred, but even to which kind to refer any one *individual*
fallacy.' Thus, so intimately are our intellectual operations blended,
that it is often extremely difficult to decide whether a mistake be
mainly due to defective observation or erroneous reasoning.

[49] See pp. 6, 23, 127–8, 131–4.

causes, the causes may either simply contribute towards the production of the total result, though one only would produce some portion of it, or they may all be essential to the production of any result whatever. It would be convenient if, in the former case, we could speak of the causes as *joint causes*, in the latter as *joint conditions*, but to do so would perhaps be too great an innovation on established language.

(*a*) An instance of supposing that a phenomenon is entirely due to one cause, when it seems in reality to be only partially due to it, is furnished by the prevalent notion that the heart is the sole cause of the circulation of the blood.

'What is it,' says Mr. Lewes[50], 'which causes the blood to circulate ? "The heart," answers an unhesitating reader. That the heart pumps blood incessantly into the arteries, and that this pumping must drive the stream onwards with great force, there is no doubt ; but, although the most powerful agent in the circulation, the heart is not the sole agent; and the more we study this difficult question, the more our doubts gather round the explanation.'

'Let a few of the difficulties be stated. There have been cases of men and animals born without a heart; these "acardiac monsters" did not live, indeed could not live ; but they had grown and developed in the womb, and consequently their blood must have circulated. In most of these cases there has been a twin embryo, which was perfect ; and the circulation in both was formerly attributed to the heart of the one ; but it has been fully established that this is not the case. Further, Dr. Carpenter reminds us that "it has occa-

[50] *Physiology of Common Life*, vol. i. p. 322.

sionally been noticed that a degeneration in the structure of the heart has taken place, during life, to such an extent that scarcely any muscular tissue could at last be detected in it, but without any such interruption to the circulation as must have been anticipated if this organ furnishes the sole impelling force." On the other hand, an influence acting on the capillaries will give a complete check to the action of the heart, although that organ is itself perfectly healthy and vigorous.'

Mr. Lewes then proceeds to discuss the subject at greater length, but the above quotation is sufficient for my purpose.

A familiar instance of this error occurs in the vulgar notion that the mean annual temperature of a place is exclusively determined by its latitude. The reader need hardly be told that in this case there are many other causes at work, namely, elevation, distance from the sea, proximity of mountain chains, and the like.

When a number of causes contribute towards the total effect, it is plain that, as in the last instance, they may operate in the way of modifying, counteracting, or even frustrating [51] each other's influence. This is a consideration which it is often of the utmost importance to bear in mind, as will be obvious from the following examples, extracted, the former from Dr. Paris' *Pharmacologia* [52], the latter from Sir G. C. Lewis' *Methods of Observation and Reasoning in Politics* [53].

[51] We sometimes speak of causes ' wholly or partially counteracting each other.' It would be an advantage if we could appropriate the word *frustration* to express complete counteraction.

[52] P. 498.   [53] Vol. i. p. 386.

' In ordering saline draughts as vehicles for active medicines, it is very important that they should be rendered perfectly neutral; the effect of a predominating acid or alkali may produce decompositions fatal to the efficacy of the remedy, as the practitioner will fully understand by a reference to the *Acetate of Ammonia* and other preparations in the Table of Incompatibles. In prescribing them to be taken in a state of effervescence, we must consider whether the disengaged carbonic acid may not invalidate the powers of the remedies simultaneously given with them. I should certainly recommend such a form to be avoided, in all cases where a salt of lead had been administered, for the carbonic acid retained in the stomach might probably convert it into a *carbonate.*'

' But it is to be borne in mind that, in estimating negative instances, due allowance must be made for the occasional *frustration of causes.* . . . . For example : it might be argued, from the occurrence of several cases in which the absence of high import duties and of commercial restrictions was accompanied with abundance and cheapness of commodities, that the former was the cause of the latter. Certain instances might then occur, in which the former existed without the latter ; but each of these exceptional cases might be accounted for, by showing that there was a special circumstance, such as a deficient supply, or interruption of intercourse by war or blockade, which partially obstructed, and for a time suspended, the operation of the former cause. Again : it might be shown, by the evidence of facts, that the operation of a new law had been generally beneficial, with the exception of certain districts, where its enforcement had been prevented or retarded by certain peculiar and accidental circumstances. Exceptions of this kind, which admit of an adequate special explanation, serve rather to confirm the general inference than to weaken it ; inasmuch as they raise the presumption that, but for the partial obstruction to the cause, it would have operated in these as in the other instances where no obstructions existed [54].'

' It is probably from observing this case of the problem of causation, that the popular error has arisen of supposing that a rule is sometimes proved by its exceptions. Every exception to a general proposition must, in so far as it is an exception, detract from the application of the proposition, and consequently disprove [or rather go towards disproving] it. Thus, if it were asserted that all cloven-footed animals ruminate, this assertion certainly would receive no confirmation from the fact, that certain cloven-footed animals—such as the hog—do not ruminate. If, however, the exception, as in the case which we have been examining, admitted of a peculiar explanation, and it could be shown that the *nisus* or tendency of the cause was the same in the exceptional as in the other instances, but that in the former it was counteracted and overcome, while in the latter it was not—then the exception may be said not to invalidate, but rather to confirm the rule.'

The above passage is noteworthy, as furnishing a good comment on the maxim, *Exceptio probat regulam*, a maxim which is, of course, only applicable where the exceptions are apparent, and where they admit of explanation in conformity with the rule.

(β) That every event depends upon the concurrence of a number of causes, positive and negative, or, as they are often called, conditions, has already been pointed out (Chap. I. pp. 13–16). Thus, the burning of a fire

[54] Sir G. C. Lewis' *Methods of Observation and Reasoning in Politics*, vol. i. p. 386.

depends not only on the application of a lighted match and the supply of fuel, but also on the presence of atmospheric air, or rather of the oxygen which it contains, though, from the universal presence of air, we are less apt to think of the latter cause than of the former ones. The importance, however, of not overlooking this consideration is shown by the extent to which we can augment the temperature by constantly bringing fresh currents of air into contact with any heated mass, as well as by the similar and familiar phenomenon of the increased brightness with which a fire burns on a frosty day, owing to the better draught.

The importance of bearing in mind that an event depends upon a concurrence of causes may be further illustrated by the boiling-point of water. The point at which water (by which I mean pure water) boils depends slightly on the nature of the vessel, but mainly upon two causes or conditions, the temperature of the water and the pressure of the atmosphere. Now, as the latter varies at different heights and in different states of weather, water does not always boil at the same temperature, the boiling-point being, as a rule, diminished by 1° for every 590 feet that we ascend, so that, whereas at the sea level water boils at about 212° Fahrenheit, on the top of Mont Blanc it boils at about 185°. It is obvious that any one, not bearing in mind this fact, might be exposed to the greatest practical inconveniences.

The following quotations from Dr. Paris' *Pharmacologia* will furnish a sufficient illustration of the importance of this consideration and of the errors which may result from neglecting it.

'In some cases of irritability of stomach, the addition of a small quantity of opium will impart efficacy to a remedy otherwise inert; an emetic will often thus be rendered more active, as I have frequently witnessed in my practice. In some states of mania, and affections of the brain, emetics will wholly fail, unless the stomach be previously influenced and prepared by a narcotic. I have often also found that the system has been rendered more susceptible of the influence of mercury by its combination with antimony and opium. So, again, when the system is in that condition which is indicated by a hot and dry skin, squill will fail in exciting expectoration; but administer it in conjunction with ammonia, and in some cases with *Antimonial Wine* and a *saline draught*, and its operation will be promoted. As a diuretic, *Squill* is by no means active, when singly administered, but *Calomel*, or some mercurial, when in combination with it, appears to direct its influence to the kidneys, and in some unknown manner to render these organs more susceptible of its influence[55].'

'It has been determined by the most ample experience that substances will produce effects upon the living system, when presented in a state of simple mechanical mixture, very different from those which the same medicinal ingredients will occasion when they are combined by the agency of chemical affinity. To illustrate this by a simple case,—a body suspended in a mixture in the form of a powder, will act very differently if held in solution by a fluid. The relative effects of alcohol in the form of what is termed "*spirit*," and in that of wine, may be explained upon the same principle; in the former case it is in a state of mixture, in the latter in

[55] *Pharmacologia*, p. 388.

that of combination. It has been demonstrated, beyond all doubt, that a bottle of port, madeira, or sherry, actually contains as much alcohol as exists in a pint of brandy ; and yet how different the effect!—a fact which affords a very striking illustration of the extraordinary powers of chemical combination in modifying the activity of substances upon the living system[56].'

'It has been very generally supposed that substances, whose application does not produce any sensible action upon the healthy system, cannot possess medicinal energy; and, on the contrary, that those which occasion an obvious effect must necessarily prove active in the cure or palliation of disease. To this general proposition, under certain limitations and restrictions, we may perhaps venture to yield our assent ; but it cannot be too early, nor too forcibly impressed upon the mind of the young practitioner, that *medicines are, for the most part, but relative agents, producing their effects in reference only to the state of the living frame.* We must, therefore, concur with Sir Gilbert Blane in stating that the virtues of medicines cannot be fairly essayed, nor beneficially ascertained, by trying their effects on sound subjects, because that particular morbid condition does not exist which they may be exclusively calculated to remove ; thus, in a robust state of the body, the effects of *steel*, in commendation of which, in certain diseases, professional opinion is unanimous, may be wholly imperceptible. Bitter tonics, also, may either prove entirely inert, or they may give strength, relax the bowels, or induce constipation, according to the particular condition of the patient to whom they are administered ; so again, in a healthy state of the stomach, a few grains of soda or magnesia will not occasion the least sensible effect, but, where that organ is infested with a morbid acid, immediate relief will follow the ingestion of the one, and purgation that

[56] *Pharmacologia*, pp. 426, 427.

of the other. By not reasoning upon such facts, physicians have, in my opinion, very unphilosophically advanced to conclusions respecting the inefficacy of certain agents. They have administered particular preparations in large doses, and, not having observed any visible effects, have at once denounced them as inert. I might allude, for instance, to the *tris-nitrate of bismuth,* a substance which, however powerless in health, I am well satisfied, from ample experience, is highly efficacious in controlling certain morbid states of the stomach. Dr. Robertson has well observed that disease calls forth the powers, and modifies the influence of medicines. That which agitates the calm of health may soothe the irritation of illness, and that, which without opposition is inert, may act powerfully where it meets with an opponent. Experiments should be made on the sick, in order to determine how the sick will be affected, and nothing should be pronounced feeble, merely because it has done nothing where there was nothing to be done[57].'

To adduce one more illustration: insanity, though sometimes due to a number of causes, each one of which simply contributes to and augments the affection, which would still exist, though in a weaker degree, even if some of them were absent, appears at other times to be the joint result of a number of causes, the presence of every one of which seems to be essential to the production of any effect so definite as to deserve the name of mental derangement. The train, in these cases, appears to be laid by a number of precedent circumstances, and the addition of some one other circumstance seems to be the spark which produces the conflagration.

[57] *Pharmacologia,* pp. 133, 134.

'When we are told,' says Dr. Maudsley[58], 'that a man has
become deranged from anxiety or grief, we have learned very
little if we rest content with that. How does it happen that
another man, subjected to an exactly similar cause of grief,
does not go mad? It is certain that the entire causes cannot
be the same where the effects are so different; and what we
want to have laid bare is the conspiracy of conditions,
internal and external, by which a mental shock, inoperative
in one case, has had such serious consequences in another.
A complete biographical account of the individual, not neglecting the consideration of his hereditary antecedents, would
alone suffice to set forth distinctly the causation of his
insanity. If all the circumstances, internal and external,
were duly scanned and weighed, it would be found that there
is no accident in madness; the disease, whatever form it
might take, by whatsoever complex concurrence of conditions, or by how many successive links of causation, it
might be generated, would be traceable as the inevitable
consequence of certain antecedents, as plainly as the explosion of gunpowder may be traced to its causes, whether
the train of events of which it is the issue be long or short.
The germs of insanity are sometimes latent in the foundations of the character, and the final outbreak is perhaps the
explosion of a long train of antecedent preparations.'

(3) The phenomena of insanity also furnish a good
illustration of the next source of error, the mistaking of
joint effects for cause and effect. In this, as in many
other diseases, symptoms are often mistaken for causes.
Thus, it is not uncommon to hear violent religious excitement or inordinate grief adduced as causes of
insanity, whereas these are probably merely incipient

[58] *Physiology and Pathology of Mind*, Part. II. ch. i. p. 225.

symptoms, due, in the vast majority of cases, to precisely the same combination of physical and mental causes, which, when they operate with greater intensity, ultimately issue in definite and unmistakable insanity.

We have an instructive instance of the same error in some of the speculations respecting the origin of fevers. In Abdominal Typhus (the so-called Typhoïd or Enteric Fever of the English Physicians) the febrile symptoms (Pyrexia, Erethism, &c.) have been ascribed to certain lesions of the glandular structures of the intestines; but a wider observation has shown that the other symptoms often precede by some time the formation of the lesions, and that the fever may even run a fatal course, though it may be impossible, in a post-mortem examination, to detect the specific lesions in question. Practically, the correction of this and similar errors is of great importance, as much mischief may be done, and much time may be lost, by a mode of treatment which, through mistaking symptoms for causes, or co-effects for cause and effect, addresses itself only to the consequences of the malady, and leaves the real source of evil unattacked.

The following anecdote, told by Dr. Paris, affords an amusing illustration of the extent to which the ignorant, in reasoning on cause and effect, may be deceived by an invariable, or even frequent, concurrence of events.

'It should,' says he[59], 'be kept in mind, that two events may arise from a common cause, and be co-existent, and yet have not the most remote analogy to, or dependence upon,

[59] *Pharmacologia,* p. 89.

each other. It was a general belief at St. Kilda, that the arrival of a ship gave all the inhabitants *colds*. Dr. John Campbell took a great deal of pains to ascertain the fact, and to explain it as the effect of effluvia arising from human bodies ; the simple truth, however, was that the situation of St. Kilda renders a north-east wind indispensably necessary before a stranger can land,—the wind, not the stranger, occasioned the epidemic.'

In speculations on the history of language, languages, which recent investigation has shown to be related collaterally, were by older philologists erroneously regarded as standing to each other in the relation of parent and child. I extract from Professor Max Müller's *Lectures on the Science of Language*[60] the following illustration, which will already be familiar to many of my readers :—

'A glance at the modern history of language will make this clearer. There never could be any doubt that the so-called Romance languages, Italian, Wallachian, Provençal[61], French, Spanish, and Portuguese, were closely related to each other. Everybody could see that they were all derived from Latin. But one of the most distinguished French scholars, Raynouard, who has done more for the history of the Romance languages and literature than any one else, maintained that Provençal only was the daughter of Latin; whereas French, Italian, Spanish, and Portuguese were

[60] First Series. Lecture V.

[61] The exact relationship of French to Provençal may be represented thus: the Peasant Latin became in the South of France the Langue d'Oc (or Provençal), and in the North the Langue d'Oil, of which the French (or the dialect of the Isle de France) was the principal dialect, and has in its modern form become the language of the nation. See Brachet's *Historical Grammar* (Dr. Kitchin's Translation), p. 18 ; 7th ed. pp. 22–3.

the daughters of Provençal. He maintained that Latin passed, from the seventh to the ninth century, through an intermediate stage, which he called Langue Romane, and which he endeavoured to prove was the same as the Provençal of Southern France, the language of the Troubadours. According to him, it was only after Latin had passed through this uniform metamorphosis, represented by the Langue Romane or Provençal, that it became broken up into the various Romance dialects of Italy, France, Spain, and Portugal. This theory, which was vigorously attacked by August Wilhelm von Schlegel, and afterwards minutely criticised by Sir Cornewall Lewis, can only be refuted by a comparison of the Provençal grammar with that of the other Romance dialects. And here, if you take the auxiliary verb *to be*, and compare its forms in Provençal and French, you will see at once that, on several points, French has preserved the original Latin forms in a more primitive state than Provençal, and that, therefore, it is impossible to classify French as the daughter of Provençal, and as the granddaughter of Latin. We have in Provençal :—

sem, corresponding to the French *nous sommes*,
etz   ,,   *vous êtes*,
son   ,,   *ils sont*,

and it would be a grammatical miracle if crippled forms, such as *sem*, *etz*, and *son*, had been changed back again into the more healthy, more primitive, more Latin, *sommes*, *êtes*, *sont; sumus, estis, sunt.*

Let us apply the same test to Sanskrit, Greek, and Latin ; and we shall see how their mutual genealogical position is equally determined by a comparison of their grammatical forms. It is as impossible to derive Latin from Greek, or Greek from Sanskrit, as it is to treat French as a modification of Provençal. Keeping to the auxiliary verb *to be*, we find that *I am* is in

Sanskrit	Greek	Lithuanian
asmi	*esmi*	*esmi.*

The root is *as*, the termination *mi*.

Now, the termination of the second person is *si*, which together with *as*, or *es*, would make

as-si	*es-si*	*es-si.*

But here Sanskrit, as far back as its history can be traced, has reduced *assi* to *asi;* and it would be impossible to suppose that the perfect, or, as they are sometimes called, organic, forms in Greek and Lithuanian, *es-si*, could first have passed through the mutilated state of the Sanskrit *asi*.

The third person is the same in Sanskrit, Greek, and Lithuanian, *as-ti* or *es-ti;* and, with the loss of the final *i*, we recognise the Latin *est*, Gothic *ist*, and Russian *est'*.

The same auxiliary verb can be made to furnish sufficient proof that Latin never could have passed through the Greek, or what used to be called the Pelasgic stage, but that both are independent modifications of the same original language. In the singular, Latin is less primitive than Greek; for *sum* stands for *es-um*, *es* for *es-is*, *est* for *es-ti*. In the first person plural, too, *sumus* stands for *es-umus*, the Greek *es-mes*, the Sanskrit *'smas*. The second person *es-tis* is equal to Greek *es-te*, and more primitive than Sanskrit *stha*. But in the third person plural Latin is more primitive than Greek. The regular form would be *as-anti;* this, in Sanskrit, is changed into *santi*. In Greek, the initial *s* is dropped, and the Æolic *enti* is finally reduced to *eisi*. The Latin, on the contrary, has kept the radical *s*, and it would be perfectly impossible to derive the Latin *sunt* from the Greek *eisi*.'

(4) A not uncommon source of error is the confusion of the proximate with the primary or remote cause of a phenomenon. To be on our guard against this error is often of the utmost practical importance : for the remo-

val of the proximate cause may only temporarily remove the effect, and the primary cause may, after a time, reproduce it; or, again, the removal of the primary cause may still leave the proximate cause in full action. This error is well exemplified in Mr. Lewes' account of Thirst.

'The sensation of Thirst is not merely a sensation dependent on a deficiency of liquid in the system, but a local sensation dependent on a local disturbance : the more water these men (the prisoners confined in the Black Hole at Calcutta) drank, the more dreadful seemed their thirst; and the mere sight of water rendered the sensation, which before was endurable, quite intolerable. The *increase* of the sensation following a *supply* of water, would be wholly inexplicable to those who maintain that the proximate cause of Thirst is deficiency of liquid; but is not wholly inexplicable, if we regard the deficiency as the primary, not the proximate cause : for this primary cause having set up a feverish condition in the mouth and throat, that condition would continue after the original cause had ceased to exist. The stimulus of cold water is only a momentary relief in this case, and exaggerates the sensation by stimulating a greater flow of blood to the parts. If, instead of cold water, a little lukewarm tea, or milk-and-water, had been drunk, permanent relief would have been attained; or if, instead of cold water, a lump of ice had been taken into the mouth, and allowed to melt there, the effect would have been very different—a transitory application of cold increasing the flow of blood, a continuous application driving it away.

'We must not, however, forget that, although, where a deficiency of liquid has occasioned a feverish condition of the mouth and throat, no supply of cold liquid will at once remove that condition, the relief of the Systemic sensation not immediately producing relief of the local sensation, nevertheless, so long as the system is in need of liquid, the feeling of thirst must continue. Claude Bernard observed that a dog which had an opening in its stomach drank unceasingly because the water ran out as fast as it was swallowed; in vain the water moistened mouth and throat on its way to the stomach. Thirst was not appeased because the water was not absorbed. The dog drank till fatigue forced it to pause, and a few minutes afterwards recommenced the same hopeless toil; but no sooner was the opening closed, and the water retained in the stomach, from whence it was absorbed into the system, than thirst quickly vanished [62].'

In studying the history of a language, it is often most important to bear in mind that words ultimately derived from one language are proximately derived through the medium of another. Thus, there will occur to the reader numberless English words which have been derived from the Latin through the French, as, for instance, *judge, noble, emperor, governor, prince.* And, to quote M. Brachet :—

'When Jerome translated the Old Testament into Latin, he incorporated into his version certain Hebrew words which had no Latin equivalents, as seraphim, Gehenna, pascha, &c.; from Latin they passed at a later time into French (*séraphin, gêne, pâque*). But they entered French from the Latin, not from the Hebrew. The same is the case with the Arabic; its relations with French have been purely accidental. To say nothing of those words which express oriental things, such as *Alcoran, bey, cadi, caravane, derviche, firman, janissaire*, &c., which were brought into the west by travellers, the French language received, in the middle ages, many Arabic words from another source: the Crusades, the

[62] Lewes' *Physiology of Common Life*, vol. i. pp. 45–47.

scientific greatness of the Arabians, the study of oriental philosophies, much followed in France between the twelfth and fourteenth centuries, enriched the vocabulary of the language with many words belonging to the three sciences which the Arabians cultivated successfully: in astronomy it gave such words as *azimuth, nadir, zénith;* in alchemy, *alcali, alcool, alambic, alchimie, élixir, sirop;* in mathematics, *algèbre, zéro, chiffre.* But even so these words did not come directly from Arabic to French; they passed through the hands of the scientific Latin of the middle ages. In fact, the oriental languages have had little or no popular or direct influence on French[63].'

The non-recognition of these intermediate channels, through which the words of one language have been introduced into another, has often led to the most erroneous theories as to the connexion of languages or the relations subsisting between the people speaking them. Thus, it was once a favourite theory that all languages are derived from Hebrew, and the occurrence in different languages of the same words has often, without any other ground, been regarded as a proof of the connexion of the most diverse races.

I add an example from the science of Political Economy. It has often been supposed that high prices produce high wages. A sudden rise in the price of any particular class of commodities may lead, by a desire on the part of the producers to increase the supply, and by a consequent increase in the demand for labour in that particular department, to a temporary rise in wages.

[63] *Historical Grammar*, Translation, p. 22, note 2; 7th ed. p. 27, note 2.

But a rise in prices produces no permanent rise in wages, unless it leads to an increased accumulation of capital, that is, an augmentation of the fund available for the further production of wealth and, consequently, for the payment of wages[64]. Here the rise in prices is the remote or primary, and the increased accumulation of capital is the proximate, cause of the phenomenon; but, as counteracting causes, such as reckless speculation or the adoption of a more luxurious style of living on the part of the capitalists, may prevent the rise in prices from being followed by an increased accumulation of capital, it is often of great importance to distinguish the two.

I have, thus far, discussed those errors which originate in overlooking the presence of some third circumstance. But, even when all the circumstances except the cause and effect (or what we suppose to be such) have been eliminated, we may still commit an error, either from mistaking the cause for the effect, or from neglecting to take account of their mutual action and reaction and being thus led erroneously to assign to one of the two exclusively the whole share in the production of the ultimate effect.

(5) The importance of not overlooking this latter source of error is well illustrated by the following remarks of Sir G. C. Lewis[65] :—

[64] See Mill's *Political Economy*, Bk. II. ch. xi. § 2.
[65] *On Methods of Observation and Reasoning in Politics*, vol. i. p. 375.

'An additional source of error in determining political causation is likewise to be found in the *mutuality of cause and effect.* It happens sometimes that, when a relation of causation is established between two facts, it is hard to decide which, in the given case, is the cause and which the effect, because they act and re-act upon each other, each phenomenon being in turn cause and effect. Thus, habits of industry may produce wealth; while the acquisition of wealth may promote industry: again, habits of study may sharpen the understanding, and the increased acuteness of the understanding may afterwards increase the appetite for study. So an excess of population may, by impoverishing the labouring classes, be the cause of their living in bad dwellings; and, again, bad dwellings, by deteriorating the moral habits of the poor, may stimulate population. The general intelligence and good sense of the people may promote its good government, and the goodness of the government may, in its turn, increase the intelligence of the people, and contribute to the formation of sound opinions among them. Drunkenness is in general the consequence of a low degree of intelligence, as may be observed both among savages and in civilized countries. But, in return, a habit of drunkenness prevents the cultivation of the intellect, and strengthens the cause out of which it grows. As Plato remarks, education improves nature, and nature facilitates education. National character, again, is both effect and cause: it re-acts on the circumstances from which it arises. The national peculiarities of a people, its race, physical structure, climate, territory, &c., form originally a certain character, which tends to create certain institutions, political and domestic, in harmony with that character. These institutions strengthen, perpetuate, and reproduce the character out of which they grew, and so on in succession, each new effect becoming, in its turn, a new cause. Thus a brave, energetic, restless nation, exposed to attack from neighbours,

organises military institutions : these institutions promote
and maintain a warlike spirit : this warlike spirit, again,
assists the development of the military organisation, and
it is further promoted by territorial conquests and success in
war, which may be its result—each successive effect thus
adding to the cause out of which it sprung.'

The difference between the calculated and observed
velocities of sound (already noticed[66]) furnishes another
illustration of the importance of attending to the mutual
action of cause and effect. The wave of sound, in its
passage through the air, developes heat by compression,
and this heat, by augmenting the elasticity of the air,
increases, in turn, the velocity with which the sound
is transmitted. Thus the effect re-acts upon, and pro-
motes the operation of, the original cause. It was from
overlooking this fact that Newton's calculation of the
velocity of sound fell short of the observed velocity by
about one-sixth of the actual rate.

Malthus' speculations on the increase of population
illustrate another form of the same error. He found
that, in many cases, population increased faster than
food increased. He inferred that this increase of population once begun would continue under all circumstances ; and that therefore a time was at hand, in many
countries, when the bulk of the people would be reduced
almost to a state of starvation. He did not observe
that, in this case, the effect re-acts upon the cause ; not,
however, in the way of promoting but of retarding its

[66] Pp. 181–2.

operation. The tendency of an increase of population is certainly to diminish the supply of food; but, in attempting to forecast the ultimate result of this tendency, Malthus did not take sufficient account of the fact that the diminution in the supply of food has, in its turn, a tendency to arrest the increase of population.

Instances of the tendency of an effect to re-act upon its cause, in the way of diminishing its intensity, are very frequent in human affairs. Thus, when a man discovers that he is labouring under a disease, the additional prudence which he is induced to exercise will often not only arrest or retard the progress of the disease, but lead to the prolongation of his life beyond the usual term. Again, when a deficiency of sanitary arrangements has led to an increased mortality or the outbreak of a pestilence, the attention thus directed to the noxious influences at work will often result in their removal, or, at least, in some considerable alleviation of them. It is plain that, in speculating on the future, these are considerations which ought not to be left out of account.

(6) We may invert cause and effect, mistaking one for the other. This error is not infrequent in historical speculations, as, for instance, when some great event, such as the religious reformation of the sixteenth century, or the French Revolution, is assigned as the cause of a general change of opinion or of certain mental and social habits, whereas, in reality, the gradual, and often unobserved, operation of this change has been the

cause, and not the effect, of the historical event. In a case of this kind, however, the event may, in turn, have intensified, and, perhaps, given the sanction of authority to, the causes which produced it.

Again, a particular form of government, monarchical, aristocratical, democratical, or the like, is often assigned as the cause of certain peculiarities of social feeling or national character, whereas it would probably be far more correct to regard the form of government as due, in the first instance, to these peculiarities, though it, in turn, may have intensified the causes to which it was originally due.

In meteorological speculations it has been questioned whether the electrical phenomenon of lightning is the cause or effect of the sudden precipitations of rain and hail which it generally accompanies. Sir John Herschel (in opposition to the ordinary opinion [67]) maintains that it is the effect, and argues thus :—

'Whatever may be the state of the ultimate molecules of vapour, it seems impossible but that when a great multitude of them lose their vaporous state by cold, and coalesce into a drop or snow spangle, however minute, that drop will have collected and retained on its surface (according to the laws of electric equilibrium) the whole electricity of its constituent molecules, which will therefore have some finite, though very feeble tension. Now, suppose any number (1000 for instance) of such globules to coalesce, or that by successive deposition one should gradually grow to 1000 times its original *volume*. The diameter will be only 10, and the surface 100 times

[67] Herschel's *Meteorology*, §§ 135, 137.

increased. But the electric contents, being the sum of those of the elementary globules, will be increased one thousand-fold, and, being spread entirely over the surface, will have a tenfold density (*i.e.* tension).

* * * * * * *

'It will easily be seen that, when thousands of these electriferous globules again further coalesce into rain drops, a great and sudden increase of tension at their surface must take place. Their electricity, then, is enabled to spring from drop to drop, and, rushing in an instant of time from all parts of the cloud to the surface, a flash is produced. Accordingly, in thunder-storms, it is the commonest of all phenomena to find each great flash succeeded by a sudden rush of rain at such an interval of time as may be supposed to have been occupied in its descent. The sudden precipitation of large quantities of rain, and especially of hail, which is formed in a cold region, where the insulating power of the air is great, is almost sure to be accompanied with lightning, which the usual perversity of meteorologists, where electricity is in question, long persisted, and even yet persists, with few exceptions, in regarding as the cause, and not the consequence, of the precipitation.'

A question has also been raised whether the copious precipitation of rain which usually takes place in the centre of a cyclone is the cause or the effect of the cyclone. The more probable view is that the partial vacuum produced by the rain-fall, and the consequent inrush of the surrounding atmosphere, is the cause of the cyclone.

Mr. M'Lennan, in his *Primitive Marriage*, conceives that marriage by capture arose from the custom of exogamy, that is to say, from the custom which forbad marriage within the tribe. Sir John Lubbock [68], on the other hand, opposes this opinion, and regards exogamy as arising from marriage by capture, not marriage by capture from exogamy. 'Mr. M'Lennan's theory,' says he, 'seems to me quite inconsistent with the existence of tribes which have marriage by capture and yet are endogamous. The Bedouins, for instance, have unmistakeably marriage by capture, and yet the man has a right to marry his cousin, if only he be willing to give the price demanded for her.'

Professor Rogers, in his *Manual of Political Economy* [69], calls in question the received opinion on the relation between the increase of population and the cultivation of inferior soils. Though I cannot accept his position, the passage will serve as an instance of the difficulty frequently experienced in determining which of two phenomena or events is cause and which is effect.

'There is not a shadow of evidence in support of the statement that inferior lands have been occupied and cultivated as population increases. The increase of population has not preceded but followed this occupation and cultivation. It is not the pressure of population on the means of subsistence which has led men to cultivate inferior soils, but the fact that these soils being cultivated in another way, or taken into cultivation, an increased population became possible. 'How could an increased population have stimulated greater labour in agriculture, when agriculture must have supplied the means on which that increased population could have ex-

[68] *Origin of Civilization and Primitive Condition of Man*, ch. 3.
[69] p. 153.

isted[70]? To make increased population the cause of improved agriculture is to commit the absurd blunder of confounding cause and effect.'

While agreeing with the ordinary theory that the pressure of population leads, in the first instance, to the cultivation of inferior lands, I should admit that the greater area of land under cultivation, by rendering possible a larger population, reacts upon and intensifies the original cause, an increased population leading to the cultivation of fresh lands, that rendering possible a still larger population, this in turn leading to the cultivation of fresh lands, and so on, till the process is arrested by counteracting causes. If this view be correct, the ordinary theory is more justly open to the charge of neglecting to take into account the ' mutuality ' of cause and effect, noticed a few pages back, than of inverting their relation.

VI. The Argument from Analogy, as has already been stated, consists in drawing the conclusion that, because two or more phenomena resemble each other in certain observed points, they also resemble each other in certain other points beyond the range of our observation. The conditions with which such an inference, in order to be legitimate, must conform, need not be here repeated. If the conditions be not fulfilled, we may commit the error

---

[70] This question appears to ignore the fact that a population may have an insufficient supply of food, though what it does possess may be just competent to sustain life..

either of over-estimating the force of the analogy; of mistaking the direction in which it points, so as to regard an analogy which makes against a certain position as making for it, or the reverse; or, lastly, of supposing grounds of analogy to subsist where there are really none. The two former errors have been sufficiently exemplified in the chapter on Imperfect Inductions.

When we exaggerate the value of analogical evidence, or mistake the conclusion to be drawn from it, we may be led to do so either by over-rating the number of ascertained points of resemblance as compared with ascertained points of difference, or the reverse, or by miscalculating the extent of our knowledge of the phenomena. The examples referred to illustrate both sources of error. Thus, for instance, the points in which electricity resembles a fluid are obvious, while the points of difference are far less obtrusive, and, moreover, the unknown properties of electricity are probably out of all proportion to those which we know. In this case, too, when we include the consideration of heat, light, and similar agencies, the argument from analogy may be used *against*, rather than *in favour of*, the identification of electricity with a fluid.

The student need, however, hardly be reminded that an analogy which in one state of knowledge appears to be a strong one may, as knowledge advances, become extremely faint, worthless, or even positively unfavourable to the position which it was originally adduced to support.

The term False Analogy is, strictly speaking, applied

not to those cases in which we over-estimate the value
of the analogy, or mistake the direction in which the
argument points, but to those cases of analogical inference in which there exists no ground for any analogy
whatever. Two phenomena, A, B, resemble each other
in the possession of the properties a, b, c. The phenomenon A is observed also to present the property d, and
hence it is inferred as probable that the same property
is to be found also in B. Now it has already been pointed
out that if we have any special reason for supposing d to
be causally connected with any of the properties a, b, c,
the argument ceases to be analogical, and becomes inductive. But if, on the other hand, we have any special
reason for supposing that d is causally connected with
none of the properties a, b, c, there is no room for any
inference whatever. The whole force of the Argument
from Analogy consists in the *chance* of d being causally
connected with a, b, or c : if we have reason to believe
that this is the case, the argument becomes more than
analogical ; if we have reason to believe that it is not
the case, we are debarred from employing the argument
altogether. Thus, in a certain sense, the Argument from
Analogy is based on our ignorance ; it is the result of
a calculation of chances, which an accession of knowledge may invalidate, by either augmenting, diminishing,
or annihilating it. Of False Analogy, in its strict sense,
that is to say, the error of supposing that similarity or
dissimilarity in certain points is an evidence of similarity
or dissimilarity in other points, when more careful re-

flexion or observation would lead to the belief that there
is probably no connexion whatever between the observed points from which the Analogy proceeds and the
unobserved points to which it argues, instances are
extremely numerous in almost every branch of knowledge.
As this form of Fallacy is so common, I shall subjoin
several examples of it.

The following excellent illustration is quoted by Mr.
Mill from Archbishop Whately's *Rhetoric*[71]:

'It would be admitted that a great and permanent diminution in the quantity of some useful commodity, such as corn,
or coal, or iron, throughout the world, would be a serious and
lasting loss; and again that, if the fields and coal mines
yielded regularly double quantities with the same labour, we
should be so much the richer: hence it might be inferred
that, if the quantity of gold and silver in the world were
diminished one half, or were doubled, like results would
follow; the utility of these metals, for the purposes of coin,
being very great. Now there are many points of resemblance and many of difference between the precious metals
on the one hand, and corn, coal, &c. on the other: but the
important circumstance to the supposed argument is that the
utility of gold and silver (as coin, which is far the chief)
depends on their value, which is regulated by their scarcity,
or rather, to speak strictly, by the difficulty of obtaining
them; whereas, if corn and coal were ten times as abundant
(i. e. more easily obtained), a bushel of either would still be
as useful as now. But if it were twice as easy to procure
gold as it is, a sovereign would be twice as large; if only

[71] Mill's *Logic*, Bk. V. ch. v. § 6; Whately's *Rhetoric*, Part I.
ch. ii. § 7. The passage does not occur in the earlier editions of
Whately's *Rhetoric*.

half as easy, it would be of the size of a half-sovereign, and this (besides the trifling circumstance of the cheapness or dearness of gold ornaments) would be all the difference. The analogy, therefore, fails in the point essential to the argument.'

Respect for antiquity is often urged by an argument so sweeping as to assume the form of a False Analogy. 'Who are we,' it is said, 'that we should presume to think that we know better than previous generations?' Now, on many matters of fact, there can be no question that the belief of previous generations, when properly examined and sifted, must be accepted as final, inasmuch as they were contemporary, or nearly contemporary, with the original sources of information. To infer from this just and limited deference the necessity of an undiscriminating submission to the opinions of our ancestors, would be an instance of the fallacy of Inductio per Enumerationem Simplicem. But this, at least in many cases, seems not to be the nature of the argument, which appears rather to proceed on some such grounds as these: we reverence the opinions of the aged, because they have had more experience than we have had, and therefore, surely, on the same principle, we ought to accept the opinions of our ancestors, who lived in bygone generations. The point of resemblance is the fact of having been born at a period prior to ourselves, and hence it is inferred that the greater experience and the greater wisdom which are found to be concomitants of this fact in the case of many of our senior contemporaries

may also be presumed in the case of those who have
long since been dead. It, of course, escapes the notice
of those who have recourse to this argument, that the
average age of the persons living at any one time is
about the same as that of those living at any other, and
that superior wisdom is the consequence not of priority
of birth but of greater experience. Thus far, the fallacy
may be regarded as one of False Analogy, strictly so
called. But there is another consideration which turns
the edge of the argument. Experience grows with time,
each generation not only inheriting the accumulated
experience of previous generations, but adding to the
stock its own acquisitions. 'Recte enim,' says Bacon [72],

---

[72] *Novum Organum*, Lib. I. Aph. lxxxiv. In the first edition of
this work I suggested that the reference might possibly be to
Æschylus, Prometheus Vinctus, l. 981 : ἀλλ' ἐκδιδάσκει πάνθ' ὁ
γηράσκων χρόνος. Through the courtesy of the Rev. E. Marshall,
I am now enabled to supply the true reference, which is to Aulus
Gellius, *Noctes Atticæ*, Lib. XII. cap. 11 : 'Alius quidam veterum
poetarum, cujus nomen mihi nunc memoriæ non est, veritatem temporis filiam esse dixit.' It has also been pointed out to me that
' Veritas temporis filia' is the legend on the groats of Queen Mary,
which were doubtless in use in Bacon's time. The following sentences occur in the same Aphorism of the *Novum Organum :* ' De
antiquitate autem opinio, quam homines de ipsa fovent, negligens
omnino est, et vix verbo ipsi congrua. Mundi enim senium et grandæ-
vitas pro antiquitate vere habenda sunt ; quæ temporibus nostris
tribui debent, non juniori ætati mundi, qualis apud antiquos fuit. Illa
enim ætas, respectu nostri, antiqua et major ; respectu mundi ipsius,
nova et minor fuit. Atque revera quemadmodum majorem rerum
humanarum notitiam, et maturius judicium, ab homine sene expec-
tamus, quam a juvene, propter experientiam, et rerum, quas vidit, et
audivit, et cogitavit, varietatem et copiam ; eodem modo et a nostra

'. veritas temporis filia dicitur, non auctoritatis.' ' Anti-
quitas sæculi juventus mundi [73].'

Bishop Wilkins' *Discovery of a New World* contains

ætate (si vires suas nosset, et experiri et intendere vellet) majora
multo quam a priscis temporibus expectari par est; utpote ætate
mundi grandiore, et infinitis experimentis et observationibus aucta
et cumulata.' Bentham in his *Book of Fallacies*, Part I. ch. ii., and
Sydney Smith in his review of that work (*Edinburgh Review*, No.
lxxxiv, reprinted in his Collected Works), have some very apposite
and amusing remarks on this subject.

[73] *De Augmentis Scientiarum*, Lib. I. Dr. Whewell in his
Philosophy of Discovery (chap. xiii. § 4) appears to think that this
celebrated Aphorism may be traced to Giordano Bruno. 'It is
worthy of remark that a thought which is often quoted from Francis
Bacon, occurs in Bruno's *Cena di Cenere*, published in 1584; I mean,
the notion that the later times are more aged than the earlier. . In
the course of the dialogue, the Pedant, who is one of the interlocutors,
says, "In antiquity is wisdom ;" to which the Philosophical Character
replies, "If you knew what you were talking about, you would see
that your principle leads to the opposite result of that which you
wish to infer ;—I mean, that *we* are older, and have lived longer, than
our predecessors." He then proceeds to apply this thought, by
tracing the course of astronomy through the earlier astronomers up
to Copernicus.' See Wagner's edition of *Giordano Bruno's Works*,
vol. i. p. 132. In the original the passage runs thus :—' *Prudenzio.*
Sii come la si vuole, io non voglio discostarmi dal parer de gli
antichi ; per che dice il saggio : Ne l'(antiquità è la sapienza. *Teofilo.*
E soggiunge : In molti anni la prudenza. Se voi intendeste bene
quel che dite, vedreste, che dal vostro fondamento s' inferisce il con-
trario di quel che pensate : voglio dire, che noi siamo più vecchi et
abbiamo più lunga età, che i nostri predecessori.' Mr. Spedding,
however, in his edition of Bacon, questions whether Bacon intended
the aphorism as a quotation, and thinks it probable that he did not
derive it from any earlier writer.—See Ellis and Spedding's edition
of *Bacon*, vol. i. p. 458, n. 4.

the following curious extract, translated from the work of Cardinal Nicolò de Cusa *De doctâ Ignorantiâ* [74]:

'We may conjecture the inhabitants of the sun are like to the nature of that planet, more clear and bright, more intellectual than those in the moon where they are nearer to the nature of that duller planet, and those of the earth being more gross and material than either, so that these intellectual natures in the sun are more form than matter, those in the earth more matter than form, and those in the moon betwixt both. This we may guess from the fiery influence of the sun, the watery and aerous influence of the moon, so also the material heaviness of the earth. In some such manner likewise is it with the regions of the other stars; for we conjecture that none of them are without inhabitants, but that there are so many particular worlds and parts of this one universe as there are stars, which are innumerable, unless it be to Him who created all things in number.'

The analogy in this case is founded not, as in the previous instances, on points of resemblance but on points of dissimilarity. The sun, the moon, and the earth are formed of different materials, and, therefore, it is argued, their inhabitants differ in their intellectual capacities, the exaltation of intelligence rising in proportion to the 'clearness and brightness' of the globe which they inhabit. Waiving the assumptions as to the materials of which the three bodies are composed and the habitation of them all by intelligent beings, it is plain that there is no presumption in favour of the theory that the intelligence of the inhabitants stands in

[74] Wilkins' *Discovery of a New World in the Moon*, p. 128; Cusanus, *De doctâ Ignorantiâ*, Lib. II. ch. xii.

any relation to the material of the globe on which they live; by parity of reasoning, birds ought to be far more intelligent than men.

The following passage from Bacon's *Novum Organum*[75] furnishes a remarkable example of a combination of Confusion of Language with False Analogy: 'Sed temporibus insequentibus, ex inundatione Barbarorum in imperium Romanum, postquam doctrina humana velut naufragium perpessa esset; tum demum philosophiæ Aristotelis et Platonis, tanquam tabulæ ex materia leviore et minus solida, per fluctus temporum servatæ sunt.' The studènt may exercise his sagacity in assigning its due share to each source of deception.

The arguments for or against the independence of colonies will often be found to rest on a False Analogy. Sometimes it is said that, under no circumstances, ought a colony to rebel against the authority of the mother-country; at other times, that, the colony having come to maturity, the time for its emancipation has arrived. In each of these cases the argument is suggested by the term 'mother-country.' Now the relations of the child to the parent are mainly determined by natural affection, by early associations, by gratitude for favours received, and frequently by the fact that, while the child is gradually approaching to the prime of life, the parent is gradually receding from it. Similar circumstances, though to a far weaker degree, may undoubtedly determine the relations of a colony to its 'mother-country,'

[75] Lib. I. Aph. lxxvii.

as, for instance, sympathy of race, the associations of many of the colonists with their early home, gratitude for assistance received at the foundation of the colony or during the earlier years of its existence, the growing prosperity of the colony or the waning power of the 'mother-country.' But, in addition to the fact that there are many cases in which these circumstances or some of them do not exist, or in which they exist only to the slightest extent, it must be plain, on reflexion, that the justice or injustice, the expediency or inexpediency, of separation from the mother-country or of repudiation by it must often be settled by considerations totally distinct from these, and such as receive no elucidation whatever from the relations between parent and child.

The illusion, originating in a false analogy, that every community must, like every individual man, pass through the three stages of growth, vigour, and decay, is thus exposed by Sir G. C. Lewis [76]:

[76] *Methods of Observation and Reasoning in Politics*, vol. ii. p. 438. The Rev. E. H. Hansell has pointed out to me a striking passage in Burke's ' Letters on a Regicide Peace,' in which Sir G. C. Lewis' notice of this fallacy is anticipated : ' I am not quite of the mind of those speculators who seem assured that necessarily, and by the constitution of things, all states have the same periods of infancy, manhood, and decrepitude, that are found in the individuals who compose them. Parallels of this sort rather furnish similitudes to illustrate or to adorn, than supply analogies from whence to reason. The objects which are attempted to be forced into an analogy are not found in the same classes of existence. Individuals are physical beings, subject to laws universal and invariable. The immediate cause acting in these laws may be obscure : the general results are subjects of certain calculation. But commonwealths are not physical

'From what has been already said, it follows that the comparison which is sometimes instituted between the progress of a community and the life of a man fails in essentials, and is therefore misleading. Both a man and a community, indeed, advance from small beginnings to a state of maturity: but a man has an allotted term of life, and a culminating point from which he descends ; whereas a community has no limited course to run ; it has no necessary period of decline and decay, similar to the old age of a man ; its national existence does not necessarily cease within a certain time. Nations, as compared with other nations, have periods of prosperity and power; but even these periods often ebb and flow, and when a civilised nation loses its pre-eminence—as Italy in the nineteenth, as compared with Italy in the fourteenth and sixteenth centuries—it does not necessarily lose its civilisation. A political community is renewed by the perpetual succession of its members ; new births, immigrations, and new adoptions of citizens, keep the political body in a state of continuous youth. No such process as this takes place in an individual man. If he loses a limb, it is not replaced by a fresh growth. The effects of disease are but partially repaired ; all the bodily and mental functions are gradually enfeebled, as life is prolonged, till at last decay inevitably ends in death : whereas a community might, consistently with the laws of human nature, have a duration co-extensive with that of mankind.

' The supposed analogy between the existence of a political community and the life of a man seems to have contributed to the formation of the belief in a liability to *corruption*, inherent in every society. It was a favourite doctrine among but moral essences. They are artificial combinations, and, in their proximate efficient cause, the arbitrary productions of the human mind. We are not yet acquainted with the laws which necessarily influence the stability of that kind of work made by that kind of agent.'—*Works*, vol. viii. pp. 78, 79.

some writers of the last century that every civilised community is fated to reach a period of corruption, when its healthy and natural action ceases, and it undergoes some great deterioration. The notion of an inevitable stage of corruption in a nation was, indeed, partly suggested by the commonplaces condemnatory of luxury, derived both from the classical and ecclesiastical writers; and by the more modern eulogies of savage life. So far, however, as it was founded on the inevitable periods of decay in animal and vegetable life, the comparison was delusive; for the two relations which are brought together do not correspond. The death of individuals may, indeed, be considered a necessary condition for the progress of the society, into which they enter as temporary elements. It is by the substitution of new intelligences, and of natures not hardened to old customs, for minds whose thoughts and habits have learnt to move uniformly in the same groove, that progressive changes in human affairs are effected. The decay and death of the individual, therefore, tends not only to prevent the deterioration of the society, but to promote its improvement.'

Ancient medicine was full of false analogies. I select the following examples from Dr. Paris [77]:

'An example of reasoning by false analogy is presented to us by Paracelsus, in his work *de vitâ longâ*, wherein, speaking of antimony, he exclaims, " Sicut antimonium finit aurum, sic, eâdem ratione et formâ, corpus humanum purum reddit."'

The alchemists, or some of them, appear to have imagined that the same preparation by which they hoped to convert the baser metals into gold (called metaphori-

---

[77] *Pharmacologia*, p. 64.

cally 'the healthy man') would also be effective in removing the sources of all bodily diseases. Why should not the impurities of the human body be removable by the same means as the impurities of the metals?

'They [that is, the Arabian physicians] conceived that gold was the metallic element in a state of perfect purity, and that all the other metals differed from it in proportion only to the extent of their individual contamination ; and hence the origin of the epithet *base*, as applied to such metals. This hypo-thesis explains the origin of alchemy ; but in every history we are informed that the earlier alchemists expected, by the same means that they hoped to convert the *baser* metals into gold, to produce an universal remedy, calculated to prolong indefinitely the span of human existence.

'It is difficult to imagine what connexion could exist in their ideas between the "*Philosopher's Stone*," which was to transmute metals, and a remedy which could arrest the progress of bodily infirmity : upon searching, however, into the writings of these times, it appears probable that this conceit may have originated with the alchemists from the application of false analogies, and that the error was subsequently diffused and exaggerated by a misconstruction of alchemical metaphors.'

The old maxim that 'Nature abhors a vacuum,' the curious belief, still prevalent even amongst persons of intelligence, that the weather changes with the 'changes' of the moon, the once fashionable doctrine of the 'Social Contract' or 'Original Compact,' the explanation of moral and physical facts by applying to them the conceptions of 'perfect numbers' and 'regular solids[78],' the

[78] On this subject the reader will find some very curious informa-

Pythagorean theory of the Harmony of the Spheres, the Aristotelian doctrine of the Mean, and innumerable other instances with which the student will meet in his reading, will abundantly illustrate the nature of False Analogy and its frequency in the reasoning of early speculators.

The Argument from Final Causes[79], at least that

tion in Mill's *Logic*, Bk. V. ch. v. § 6, and Whewell's *History of the Inductive Sciences*, Bk. IV. ch. iii. § 2.

[79] 'Tum vero, ad ulteriora tendens [intellectus humanus], ad proximiora recidit, videlicet ad causas finales, quæ sunt plane ex natura hominis, potius quam universi: atque ex hoc fonte philosophiam miris modis corruperunt.'—Bacon, *Nov. Org.* Lib. I. Aph. xlviii. To prevent misconception, I may state that I am far from denying that the Argument from Final Causes, if it take sufficient account of the evolution of organisms and their power of adapting themselves to external circumstances, and if it be based on the contemplation of Nature as a whole, instead of on that of individual objects, may admit of being stated in such a form as to occupy once more an important position in any scheme of Natural Theology. Bearing in mind these qualifications, it may be perfectly legitimate to speak, with reference to the universe at large, of design and a designer, whatever may have been the agency, and however mysterious and prolonged the process, by which an intelligent Creator may have worked. Theories of evolution may be so stated as not to impair, but indefinitely to exalt, our ideas of the power, wisdom, and benevolence of the Being in whom Nature had its source.

The student will find a very temperate statement of the prevalent theory, together with much useful information on the literature of the subject, in Sir H. Acland's *Harveian Oration* for 1865. On the other side he may, with most advantage, consult the works of Mr. Herbert Spencer and Mr. Darwin. I have discussed the subject, with special reference to the views of Bacon, in the Introduction to my edition of the *Novum Organum*, § 10.

extreme form of it which assumes that every natural organism was specially designed to subserve some special object, and fashioned, once for all, in immediate reference to that object, appears ultimately to repose on a False Analogy. God or Nature (for both terms are used) is assimilated to a human artificer, and the argument appears to rest on the assumption that the motives, conceptions, and contrivances of the one may be regarded as similar to those of the other. 'Nature does nothing in vain.' 'Nature always acts for the best.' 'Everything is designed for some good purpose.' These and similar maxims express the general principle on which the argument rests. Of its application to special cases we may take the following examples.

The instances given by Bacon, in the *Advancement of Learning* and the *De Augmentis*[80], when protesting

---

[80] *Advancement of Learning,* Bk. II. (Ellis and Spedding's edition, vol. iii. p. 358). Cp. *De Augmentis,* iii. 4. It should be noticed, however, that Bacon allows the use of Final Causes in what he calls 'Metaphysic.' Of the foregoing instances, 'and the like,' he says that they are 'well enquired and collected in Metaphysic; but in Physic they are impertinent.' And again: 'Not because these final causes are not true, and worthy to be enquired, being kept within their own province; but because their excursions into the limits of physical causes hath bred a vastness and solitude in that track.' What Bacon appears to mean (and the distinction is important) is that, in extra-physical speculations, as are those of Natural Theology (for the study of 'Metaphysics,' in the ordinary sense of the term, he repudiated), we may argue from an ascertained case of adaptation to the wisdom or goodness of the Creator, but that we are not justified in assuming adaptation or design as a *datum* in physical investigation. Those who defend this use of the argument, would

against the employment of Final Causes in physical enquiries, are the following: 'The hairs of the eye-lids are for a quickset and fence about the sight; the firmness of the skins and hides of living creatures is to defend them from the extremities of heat or cold; the bones are for the columns or beams, whereupon the frames of the bodies of living creatures are built; the leaves of trees are for protecting of the fruit; the clouds are for watering of the earth; the solidness of the earth is for the station and mansion of living creatures.'

The absurd extent to which the argument may be carried by speculators who attempt to find a Final Cause for every phenomenon which falls under their cognisance, will be plain from the examples which follow. It would, however, be unjust to charge these absurdities to the account of those writers of the past generation who took a more sober, though, perhaps, an erroneous view of the argument.

In the *Timæus* of Plato [81], the construction of the

reply that many discoveries (such as, notably, Harvey's discovery of the circulation of the blood, which set out from observing the action of the valves in the veins of many parts of the body, and enquiring into their purpose) have been suggested by the idea of adaptation (which, it may be noticed, does not necessarily include the idea of design). See Acland's *Harveian Oration*, and Dugald Stewart's *Philosophy of the Human Mind*, Part II. ch. xi. (Sir W. Hamilton's edition of Stewart's Works, vol. iii. p. 335, &c.) This fact may be, and, indeed, must be, admitted with respect to physiological enquiries (however the adaptation may be accounted for), and hence Bacon's prohibition is certainly too absolute.

[81] Of Plato, Bacon says truly, that he 'ever anchoreth on that shore.'

whole universe, and specially of man, is explained on the principle of Final Causes. The following extract from Mr. Grote's *Plato* [82] will serve as a specimen of the method there employed :

'The Demiurgus, having constructed the entire Kosmos, together with the generated Gods, as well as Necessity would permit—imposed upon these Gods the task of constructing Man : the second-best of the four varieties of animals whom he considered it necessary to include in the Kosmos. He furnished to them as a basis an immortal rational soul (diluted remnant from the soul of the Kosmos) ; with which they were directed to combine two mortal souls and a body. They executed their task as well as the conditions of the problem admitted. They were obliged to include in the mortal souls pleasure and pain, audacity and fear, anger, hope, appetite, sensation, &c., with all the concomitant mischiefs. By such uncongenial adjuncts the immortal rational soul was unavoidably defiled. The constructing Gods, however, took care to defile it as little as possible. They reserved the head as a separate abode for the immortal soul : planting the mortal soul apart from it in the trunk, and establishing the neck as an isthmus of separation between the two. Again the mortal soul was itself not single but double : including two divisions, a better and a worse. The Gods kept the two parts separate ; placing the better portion in the thoracic cavity nearer to the head, and the worse portion lower down, in the abdominal cavity : the two being divided from each other by the diaphragm, built across the body as a wall· of partition : just as, in a dwelling-house, the apartments of the women are separated from those of the men. Above the diaphragm, and near to the neck, was planted the energetic, courageous, contentious soul ; so placed as to receive orders

[82] Vol. iii. pp. 272-275.

easily from the head, and to aid the rational soul in keeping under constraint the mutinous soul of appetite which was planted below the diaphragm. The immortal soul was fastened or anchored in the brain, the two mortal souls in the line of the spinal marrow continuous with the brain: which line thus formed the thread of connexion between the three. The heart was established as an outer fortress for the exercise of influence by the immortal soul over the other two. It was at the same time made the initial point of the veins, —the fountain from whence the current of blood proceeded to pass forcibly through the veins round to all parts of the body. The purpose of this arrangement is that, when the rational soul denounces some proceeding as wrong (either on the part of others without, or in the appetitive soul within), it may stimulate an ebullition of anger in the heart, and may transmit from thence its exhortations and threats through the many small blood channels to all the sensitive parts of the body; which may thus be rendered obedient everywhere to the orders of our better nature.

' In such ebullitions of anger, as well as in moments of imminent danger, the heart leaps violently, becoming overheated and distended by excess of fire. The Gods foresaw this, and provided a safeguard against it by placing the lungs close at hand with the windpipe and trachea. The lungs were constructed soft and full of internal pores and cavities like a sponge; without any blood,—but receiving, instead of blood, both the air inspired through the trachea, and the water swallowed to quench thirst. Being thus always cool, and soft like a cushion, the lungs received and deadened the violent beating and leaping of the heart; at the same time that they cooled down its excessive heat, and rendered it a more equable minister for the orders of reason.

' The third or lowest soul, of appetite and nutrition, was placed between the diaphragm and the navel. This region of the body was set apart like a manger for containing necessary food ; and the appetitive soul was tied up to it like a wild beast ; indispensable indeed for the continuance of the race, yet a troublesome adjunct, and therefore placed afar off, in order that its bellowings might disturb as little as possible the deliberations of the rational soul in the cranium for the good of the whole. The Gods knew that this appetitive soul would never listen to reason, and that it must be kept under subjection altogether by the influence of phantoms and imagery. They provided an agency for this purpose in the liver, which they placed close upon the abode of the appetitive soul. They made the liver compact, smooth, and brilliant, like a mirror reflecting images :—moreover, both sweet and bitter on occasions. The thoughts of the rational soul were thus brought within view of the appetitive soul, in the form of phantoms or images exhibited on the mirror of the liver. When the rational soul is displeased, not only images corresponding to this feeling are impressed, but the bitter properties of the liver are all called forth. It becomes crumbled, discoloured, dark, and rough ; the gall bladder is compressed ; the veins carrying the blood are blocked up, and pain as well as sickness arise. On the contrary, when the rational soul is satisfied, so as .to send forth mild and complacent inspirations,—all this bitterness of the liver is tranquillised, and all its native sweetness called forth. The whole structure becomes straight and smooth ; and the images impressed upon it are rendered propitious. It is thus through the liver, and by means of these images, that the rational soul maintains its ascendancy over the appetitive soul ; either to terrify and subdue, or to comfort and encourage it.

' Moreover, the liver was made to serve another purpose. It was selected as the seat of the prophetic agency ; which the Gods considered to be indispensable, as a refuge and aid for the irrational department of man. Though this portion of the soul had no concern with sense or reason, they would not shut it out altogether from some glimpse of truth. The revelations of prophecy were accordingly signified on the liver, for the instruction and within the easy view of the appetitive soul; and chiefly at periods when the functions of the rational soul are suspended—either during sleep, or disease, or fits of temporary extasy. For no man in his perfect senses comes under the influence of a genuine prophetic inspiration. Sense and intelligence are often required to interpret prophecies, and to determine what is meant by dreams or signs or prognostics of other kinds, but such revelations are received by men destitute of sense. To receive them, is the business of one class of men: to interpret them, that of another. It is a grave mistake, though often committed, to confound the two. It was in order to furnish prophecy to man, therefore, that the Gods devised both the structure and the place of the liver. During life, the prophetic indications are clearly marked upon it: but after death they become obscure and hard to decypher.

'The spleen was placed near the liver, corresponding to it on the left side, in order to take off from it any impure or excessive accretions or accumulations, and thus to preserve it clean and pure.'

Aristotle constantly employs this method of reasoning. Thus, in a familiar passage of the *Ethics* [83], he says that 'if it is better for men to attain happiness through their own exertions than through chance, it is reasonable to suppose that this will be the case, since everything that depends on Nature [84] is in the best possible condition.'

---

[83] *Eth. Nic.* i. 9 (5). Εἰ δ' ἐστὶν οὕτω βέλτιον ἢ διὰ τύχην εὐδαιμονεῖν, εὔλογον ἔχειν οὕτως, εἴπερ τὰ κατὰ φύσιν, ὡς οἷόν τε κάλλιστα ἔχειν, οὕτω πέφυκεν.

[84] The student will notice the transition from the Demiurgus and inferior gods of Plato to the 'Nature' of Aristotle. 'And in this,'

From his physiological works (in which the argument is most commonly employed) it will be sufficient to adduce one or two examples, which will serve also to show how a preconceived opinion may lead an author to invent false facts for the purpose of supporting his theory.

Having fixed the seat of sensation in the heart, inasmuch as it is in the centre of the body, rather than in the brain, as some philosophers had done, it was necessary to discover a special function for the brain. The necessity of discovering some function for it led to the fiction of its 'coldness,' which was supposed to counteract the heat of the heart, and so to preserve the body 'in a mean state [85].' On this account, he supposed, all animals which have blood are furnished with a brain, while bloodless animals, having little heat, require nothing to cool them, and are, therefore, without one. Moreover, in order to temper the coldness of the brain, blood is conveyed to the membrane which

says Bacon, 'Aristotle is more to be blamed than Plato, seeing that he left out the fountain of final causes, namely God, and substituted Nature for God; and took in final causes themselves rather as the lover of logic than of theology.'—*The Dignity and Advancement of Learning* (Translation of the *De Augmentis*), Bk. III. ch. iv. (Ellis and Spedding's edition, vol. iv. p. 364.)

[85] Compare the extraordinary fancy (*De Partibus Animalium*, iii. 4) that the reason why the heart, in man, inclines slightly towards the left side is that it may temper the greater coldness of that side (πρὸς τὸ ἀνισοῦν τὴν κατάψυξιν τῶν ἀριστερῶν· μάλιστα γὰρ τῶν ἄλλων ζῴων ἄνθρωπος ἔχει κατεψυγμένα τὰ ἀριστερά). It is needless to observe that the left side of man is not colder than the right; the fact is simply assumed in order to account for the position of the heart in a manner conformable with Aristotle's theories.

envelopes it by means of veins or channels. But, again, lest the heat so conveyed should injure the brain, the veins, instead of being large and few, are small and many, and the blood conveyed, instead of being copious and thick, is thin and pure[86].

'The viscera are formed out of the blood, and therefore are only found in sanguineous animals, which necessarily have a heart: for it is clear that, having blood, which is a fluid, they must have a vessel to contain it, and hence also Nature has created veins; and for these veins the origin must necessarily be one, since one, whenever possible, is better than many. The heart is the origin of the veins: this is seen in the fact that they spring from it, and do not go through it ; also they resemble it in structure. The heart has the chief position, namely, that of the centre, but more upwards than downwards, and rather in front than behind : for Nature is accustomed to seat the noblest in the noblest place, unless any stronger reason prevails : οὗ μή τι κωλύει μεῖζον[87].'

The work of Bishop Wilkins, already quoted, furnishes some curious examples of the arguments which, even within the last two hundred years, have found favour with men distinguished for their scientific attainments[88].

'Though there are some who think mountains to be a deformity to the earth, as if they were either beat up by the flood, or else cast up like so many heaps of rubbish left at the

---

[86] *De Partibus Animalium*, ii. 7. Cp. Lewes' *Aristotle*, § 164 p. 180.

[87] *De Partibus Animalium*, iii. 4. I here quote Mr. Lewes' summary, given in § 395, p. 310, of his *Aristotle*.

[88] Bishop Wilkins was one of the founders of the Royal Society, and enjoyed one of the highest scientific reputations of his time.

Creation; yet, if well considered, they will be found as much to conduce to the beauty and conveniency of the universe, as any of the other parts. Nature (saith Pliny) purposely framed them for many excellent uses: partly to tame the violence of greater rivers, to strengthen certain joints within the veins and bowels of the earth, to break the force of the sea's inundation, and for the safety of the earth's inhabitants, whether beasts or men[89].'

' I have now sufficiently proved that there are hills in the moon, and hence it may seem likely that there is also a world; for, since Providence hath some special end in all its works, certainly then these mountains were not produced in vain ; and what more probable meaning can we conceive there should be, than to make that place convenient for habitation[90] ? '

'It hath been before confirmed that there was a sphere of thick vaporous air encompassing the moon, as the first and second regions do this earth. I have now showed that thence such exhalations may proceed as do produce the comets. Now from hence it may probably follow that there may be wind also and rain, with such other meteors as are common amongst us. This consequence is so dependent that Fromondus dares not deny it, though he would (as he confesses himself); for, if the sun be able to exhale from them such fumes as may cause comets, why not such as may cause winds, why not then such also as may cause rain, since I have above showed that there is sea and land, as with us? Now rain seems to be more especially requisite for them, since it may allay the heat and scorchings of the sun, when he is over their heads. And Nature hath thus provided for those in Peru, with the other inhabitants under the line[91].'

One of the most whimsical applications of the Argu-

[89] *A Discovery of a New World in the Moon,* p. 77.
[90] Id. p. 91.   [91] Id. p. 121.

ment from Final Causes is to be found in the 'Doctrine of Signatures,' of which Dr. Paris thus speaks [92].

'But the most absurd and preposterous hypothesis that has disgraced the annals of medicine, and bestowed medicinal reputation upon substances of no intrinsic worth, is that of the " DOCTRINE OF SIGNATURES," as it has been called, which is no less than a belief that *every natural substance, which possesses any medicinal virtues, indicates, by an obvious and well-marked external character, the disease for which it is a remedy, or the object for which it should be employed.* This extraordinary monster of the fancy has been principally adopted and cherished by Paracelsus, Baptista Porta, and Crollius, although traces of its existence may certainly be discovered in very ancient authors.

* * * * * * *

'The conceit, however, did not assume the importance of a theory until the end of the fourteenth century, at which period we find several authors engaged in the support of its truth, and it will not be unamusing to offer a specimen of their sophistry : they affirm that, since man is the lord of the creation, all other creatures are designed for his use, and *therefore* that their beneficial qualities and excellences must be expressed by such characters as can be seen and understood by every one ; and as man discovers his reason by speech, and brutes their sensations by various sounds, motions, and gestures, so the vast variety and diversity of figures, colours, and consistencies, observable in inanimate creatures, is certainly designed for some wise purpose. It *must be*, in order to manifest those peculiar properties and excellences, which could not be so effectually done in any other way, not even by speech, since no language is universal. Thus, the lungs of a fox *must* be a specific for

[92] *Pharmacologia*, pp. 47–50.

asthma, *because* that animal is remarkable for its strong powers of respiration. *Turmerick* has a brilliant yellow colour, which indicates that it has the power of curing jaundice; by the same rule, *Poppies* must relieve diseases of the head; ... and the *Euphrasia* (eye-bright) acquired fame as an application in complaints of the eye, because it exhibits a black spot in its corolla resembling the pupil. In the curious work of *Chrysostom Magnenus* (Exercit. de Tabaco), we meet with a whimsical account of the *signature* of tobacco. "In the first place," says he, "the manner in which the flowers adhere to the head of the plant indicates the *infundibulum cerebri*, and *pituitary gland;* in the next place, the three membranes, of which its leaves are composed, announce their value to the stomach, which has three membranes."

'The blood-stone, the *heliotropium* of the ancients, from the occasional small specks or points of a blood-red colour exhibited on its green surface, is even at this day employed in many parts of England and Scotland to stop a bleeding from the nose; the nettle-tea continues a popular remedy for *urticaria*.

*   *   *   *   *   *   *

'It is also asserted that some substances bear the SIGNATURES of the humours, as the petals of the red rose that of the blood, and the roots of rhubarb, and the flowers. of saffron, that of the bile.

'I apprehend that John of Gaddesden, in the fourteenth century, celebrated by Chaucer, must have been directed by some remote analogy of this kind, when he ordered the son of Edward I, who was dangerously ill with the small-pox, to be wrapped in scarlet cloth, as well as all those who attended upon him, or came into his presence; and even the bed and room in which he was laid were covered with the same drapery; and so completely did it answer, say the credulous historians of that day, that the prince was cured without having so much as a single mark left upon him.'

In these and similar instances, which might be multiplied to almost any extent[93], it is plain that much is gained by the employment of the vague word Nature. Presuming that the majority of at least the more modern writers who have employed the Argument from Final Causes, if pressed to attach a definite meaning to the word Nature, would reply that they regard it, in this connexion, as only another name for God, the argument, as employed in the above and similar examples (I am not here discussing the more refined employment of it), seems to rest on the three following assumptions :—

(1) That God [or Nature] acts, not by laws, governing the evolution of natural objects, but after the manner of a human artificer, having in view some special end in the production of each object and of each separate part of it.

(2) That all objects are designed for the good of man, or, at least, of sentient or intelligent beings.

(3) That we are so well acquainted with what is, on the whole, good for ourselves, or others, or the world at large, as well as with the general plan of the universe, that we are able, in each case, to pronounce positively

[93] The following example (taken from Plutarch, *De Stoicorum Repugnantiis*, p. 1042, by Mr. Lecky, in his *History of European Morals, from Augustus to Charlemagne*, vol. ii. p. 174, note 2) is perhaps unsurpassed in absurdity: ‘Chrysippus maintained that cock-fighting was the final cause of cocks, these birds being made by Providence in order to inspire us by the example of their courage.’

on the ends which God [or Nature] proposed to himself in his constructions[94].

Of these three assumptions, the first and second are, as I conceive, based on false analogies, the first transferring to God [or Nature] the habit, observed in the human artificer, of producing each object with reference to some special end, and the second the motives which usually guide the artificer in the selection of those ends.

---

[94] The 'principle' laid down by Descartes (*De Principiis Philosophiæ*, i. 28) supplies an appropriate commentary on this assumption : 'Ita denique nullas unquam rationes circa res naturales a fine, quem Deus aut natura in iis faciendis sibi proposuit, desumemus; quia non tantum nobis debemus arrogare, ut ejus consiliorum participes nos esse putemus.'

It is interesting to compare the following extracts from Galileo's *Systema Cosmicum*, Dial. III. (Sir Thomas Salusbury's translation, pp. 333, 334) :—

'SALV. Methinks we arrogate too much to our selves, *Simplicius*, whilst we will have it that the onely care of us is the adæquate work and bound, beyond which the Divine Wisdome and Power doth or disposeth of nothing. . . . . If one should tell me that an immense space interposed between the Orbs of the Planets and the Starry Sphere, deprived of stars and idle, would be vain and useless, as likewise that so great an immensity for receipt of the fixed stars as exceeds our utmost comprehension would be superfluous, I would reply that it is rashnesse to go about to make our shallow reason judg of the Works of God, and to call vain and superfluous whatsoever thing in the Universe is not subservient to us.'

'SAGR. Say rather, and I believe you would say better, that we know not what is subservient to us; and I hold it one of the greatest vanities, yea follies, that can be in the World, to say, because I know not of what use *Jupiter* or *Saturn* are to me, that therefore these Planets are superfluous, yea more, that there are no such things *in rerum natura.*'

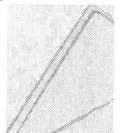

The third assumption, it need hardly be added, involves a generalisation from a very narrow range of experience to operations co-extensive with all space and all time.

Even though these various errors have been avoided, and the inductive process has been correctly performed, it is still possible, either through confusion of language, through mistaking the question at issue, or through drawing erroneous inferences in our subsequent deductions, to arrive at false conclusions. But these are considerations which properly appertain to the other branch of Logic, which is concerned with deductive reasoning.

INDEX.

Plurality of causes, 6, 23, 127–128, 131–134.

— fallacy arising from neglecting to take into account, 305.

Post hoc, ergo propter hoc, 305.

Power, question whether the idea of is involved in our conception of cause, 18–30.

Prediction, value to be attached to, 117–121.

Read, Mr. Carveth, his expression 'Vicariousness of Causes,' 127.

Reid, his criticism of Hume's account of causation, 21–25.

— his view of the nature of our conception of cause, 27.

— his view of the origin of universal beliefs, 32–33.

Social questions, the extreme difficulty attendant on their investigation, 289–291.

Species, practice of naturalists in stopping at, open to question, 80–82.

— and varieties, constant recognition of new, 82–84.

Spencer, Herbert, his view of the origin of universal beliefs, 36–37.

— referred to on the Theory of Final Causes, 342.

Statistics, conclusions based on, are instances of the application of the method of concomitant variations, 204.

Stewart, Dugald, his view of the nature of our conception of cause, 27.

Subordination of characters, principle of, 74–75.

Terminology, 92–97.

Theory, two meanings of the word, 13.

Thermotics, Science of, furnishes good examples of the Method of Difference, 155.

Type, persistency of, 83–84.

— are natural classes determined by definition or, 85–89.

Ultimate laws of nature, 225.

Uniformity of nature, law of, 5–9.

— converse does not hold true, 6.

— vaguer and more precise meanings of the expression, 9.

— universality of the belief in, 30–32 ; cp. xi–xvi.

— origin of the belief in, 7–9, 30–38.

Universal beliefs, various theories as to the origin of, 30–38.

Universal causation, law of, 4–9.

— universality of the belief in, 30–32 ; cp. xv–xvi.

— origin of the belief in, 30–38.

Variation of circumstances, importance of, 49.

Venn, Mr., referred to on a common fallacy in the calcula-

THE END.

Date Due

Lightning Source UK Ltd.
Milton Keynes UK
UKHW051950090322
399817UK00007B/80